Project Omega

The Future of Life in the Solar System

Subhash Kak and Menas C. Kafatos

गरुड

Published by
Garuda Prakashan Private Limited
Gurugram, Bharat

www.garudalife.in

First published in Bharat 2025

ISBN: 978-81-993915-5-0

Printed in Bharat

Project Omega addresses the most existential challenge that mankind has ever faced. It tackles questions that appear too difficult for many to even dare to consider them. What makes this approach unique is that it does not offer simply a technological solution (let's build a rocket ship and colonize Mars) but takes the opportunity to consider what this challenge means in terms of who we are. The book reconsiders the role of humans within the universe and seeks a deeper level of understanding. It presents concepts to provide a new framework for the future of humanity.

—Daniele C. Struppa

President of Chapman University and the Donald Bren Presidential Chair in Mathematics

This is a truly holistic book — the authors trace key developments in areas as diverse as space exploration, genetics, population trends, AI and philosophy, then weave these threads together to outline scenarios for the necessary survival of the human species outside Earth. A fascinating read!

—Ashok Narasimhan

Serial entrepreneur who has founded and run several technology companies in areas ranging from fintech to AI

Project Omega isn't just about space colonization—it's humanity's ultimate survival plan. As our Sun inevitably transforms into a life-destroying red giant, Kak and Kafatos deliver an urgent blueprint for conscious robotics and interplanetary migration. This provocative vision challenges us to transcend our earthly limitations before it's too late. Are we ready to become a multi-planetary species?

—Antonio Chella

Professor of Robotics, University of Palermo

Project Omega is a visionary and deeply thought-provoking work that challenges us, the conscious beings, to take consciousness beyond planet Earth into habitable pockets across the universe. Subhash Kak and Menas Kafatos bring together ancient wisdom, cutting-edge science, space exploration, and consciousness studies to offer a roadmap for becoming a multi-planetary species. The authors invoke action, reflection, and a profound sense of cosmic responsibility in the reader. Project Omega is both practical and deeply philosophical but offers a new framework for understanding our place in the cosmos. This is a must-read for those interested in securing real estate beyond Earth.

—Vijaya Gopal Kakani

Professor of Agronomy, University of Florida

Project Omega is a fascinating exploration of how humankind might migrate to settle *en masse* in extraterrestrial abodes over timescales of a few years from now to stellar lifespans. Multifarious scenarios driven by survival, aspirations and disruptive technologies are envisioned and examined via lenses of current and projected knowledge of objective phenomena, and subjective evolution of self-knowledge and free-will in humans and their intelligent creations. The easy-to-read book reveals subliminal connections between seemingly unrelated concepts by weaving a cogent tapestry with a cornucopia of facts, figures, hypotheses and knowledge from modern science, and wisdom from ancient philosophies and traditions. The narrative provides insights on the timeless importance of blending knowledge of the external world, and the self, to address human challenges.

—Ganpati Ramanath

John Tod Horton Professor of Materials Science and Engineering, Rensselaer Polytechnic Institute, Troy, New York

Contents

Preface

This book is about the future of life on Earth and the solar system and what can be done to safeguard it. Cosmology informs us that ultimately the Sun, *the life giver*, will become a red giant, enlarge in size and swallow up Mercury, Venus, and Earth and thus turn into a *life-taker*. Since this is estimated to happen hundreds of millions of years in the future, we have time to prepare for this eventuality. But much before that, it is conceivable that a mass extinction event like what caused the destruction of dinosaurs kills humans on Earth, or the cause of destruction may be a biological pathogen that escapes a bioweapons lab or is maliciously introduced by a madman.

We must, therefore, ensure that humans are safe at least at one other location so that, if need be, they can resettle Earth. This requires the colonization of the nearest astronomical objects, namely the Moon and Mars, to decrease the probability of human extinction and have them serve as springboard for further space exploration.

But the atmospheric conditions at the Moon and Mars are extreme. At both these places the pressure is far below at which people can survive without special suits. Shielded structures would need to be constructed to serve as pressure vessels as in a spacecraft together with complex life-support systems. Temperatures near the Moon's equator can be as high as 121°C in daylight to a low of -133°C at night. On Mars, temperatures don't fluctuate as much but they range from 20°C during day to -153°C at night. This means that most of the construction and maintenance at the settlements will have to be done by robots.

This brings us to the dual side of the project, namely that of endowing robots and other machines with as much artificial consciousness as is possible. According to a recent mathematical theorem, machines will never be conscious, but we can hope to incrementally increase the autonomy of machines to a degree that it is difficult to distinguish their performance from that of humans in a limited time window for a variety of complex tasks.

Project Omega is about the impact of evolving technology on not just life at the present time, it also deals with the issue of controlling its evolution. It will investigate challenges that humans are facing to adapt to the rapidly changing social and physical conditions brought about by modern AI technologies that are making most jobs redundant. Furthermore, there is the prospect of a worldwide population collapse which will impact society in unexpected ways. For example, China's population at the end of the century could fall as low as one-fourth of the current one, and in many European nations the natives will become a minority.

This project is also about the investigation of the biggest problem there is, namely the nature of consciousness. Behind the multitude of life forms is a single consciousness and all our knowledge of the cosmos resides within it. This aspect of the project will see a convergence of modern science with the ancient traditions of spirituality and philosophy.

The book offers the story of how we must go beyond the present consideration of life as consisting of disconnected individuals and assume our role as beings of wisdom with humility to understand our place in the cosmos. Uncertainties will drive the future, but exciting prospects await us. Science will lead to revolutionary advances in our understanding of the physical cosmos and the mystery of consciousness.

We live on an insignificant planet, however for us it is the very significant spaceship Earth. We can imagine countless planets like Earth existing in the galaxy and in the universe beyond these. If we aspire to becoming *cosmic beings*, we must begin with the colonization of the Moon and Mars. This is the first step in the long-term goal of developing technologies that will enable self-sustaining human colonies beyond Earth.

1

The Future of Life on Earth

Earth is like a speck of dust in the vast cosmos, and humans are a further smaller speck of dust on it. So how do we have the capacity to make sense of the universe? This mystery will never be resolved if we remain stuck on Earth while the Sun becomes a red giant as it is expected to in the future extinguishing all life in the solar system. Before that happens, we must carry life to all corners of the universe to preserve our collective memory. Our collective mind will guide the safeguarding of life on Earth and help us go beyond the confines of the solar system.

In the short term, we must worry about our ecosystems, the economy, social conflict, war and peace. But whatever has a beginning, has an end. For us, birth is by choice, perhaps by accident or confluence of factors, however, death is by necessity, and the same holds for Earth, other planetary systems, and stars.

The survival of our collective species is the current mega challenge. We know that we got here through *biological evolution*[1]. According to current understanding, all life on Earth shares a universal common ancestor that lived nearly 4 billion years ago. The fossil record includes a progression from early biogenic graphite to microbial mat fossils to fossilized multicellular organisms. Existing patterns of biodiversity were shaped by repeated formations of new species, changes within species, and extinction of species through the evolutionary history of life on Earth.

Dinosaurs became extinct in a moment of time, some 66 million years ago when an asteroid hit the Earth at the Yucatan Peninsula.[2] Even in the absence of a geological catastrophe, humans are at risk for a similar future. It could be a biological pathogen or a mutation that makes women unable to have children. It could even be psychological factors that lead to catastrophic population decline.

In one of the global projections from Austria's Wittgenstein Centre for Demography and Global Human Capital the human population will start collapsing by 2100 CE and by 2300 CE it could be as low as 100 million, down from the current 8 billion.[3] Even if the estimate were to be off by a factor of five, such sharp decline will have profound implications for social organization and geopolitics.

The conditions on the planets have been closely associated with the state of the Sun. In the beginning,

---

1 Emlen, D., Zimmer, C. Evolution: Making Sense of Life. W.H. Freeman (2019)

2 Cabrol, N.A. The Secret Life of the Universe. Scribner (2024)

3 Wang, B. What Disaster Happened? 2200, The World Only Has 250 Million People. Next Big Future, (2023).

nearly 4.5 billion years ago, Earth was very hot and unlivable. Eventually, water began raining from primordial clouds and as it accumulated an ocean covering the globe, Earth from being a hot planet became a wet planet. This allowed the development of life to start in the oceans and then spread to dry ground as waters receded. Life developed and flourished in different ways over millions of years of evolution. It also found ways to survive in the most extreme conditions as in Antarctica's McMurdo Dry Valleys and its Lake Vostok, which is located under two miles of the East Antarctic Ice Sheet.[4]

Water appears to have existed on the Moon at the very beginning. Mercury was always hot and devoid of any possibility of life. Venus went through runaway heating, turning its atmosphere and surface into 700 degrees with sulfuric acid rain.

For Earth, conditions became stable once the oceans formed. With the presence of water, evolution of primitive forms gave way to more complex, multicellular organisms, that is from single cells to many cells, to trilobites, to complex species and beyond. The species that arose in the ocean finally spread to solid ground. Animals and plants populated and covered Earth. Large areas of forests arose, allowing life-giving oxygen to be replenished.

Eventually, dinosaurs emerged. Non-avian dinosaurs lived between about 245 and 66 million years ago, in a time known as the Mesozoic Era. This Era is divided into three

4 Wilson, E.O. The Future of Life. Vintage Books (2002)

periods: Triassic (252 to 201 million years ago), Jurassic (201 to 145 million years ago) and Cretaceous (145 to 66 million years ago).

Supercontinent Pangaea

For Mars, water existed in liquid form and persists

as ice and permafrost. Water is also to be found on a few satellites of Jupiter and Saturn. Beyond these satellites, it is harder to guess where abundant water and enough solar energy might allow life to continue.

In the beginning, land on Earth was a huge supercontinent, Pangaea, and then it began to divide into smaller continents. The associated changes in the climate and vegetation affected how dinosaurs evolved.

The Triassic climate was relatively hot and dry, and much of the land was covered with large deserts, and there were no polar ice caps. It was in this environment that dinosaurs first arose. Reptiles tend to flourish in such hot climates because their skin is less porous than mammal skin, so it loses less water in the heat.

Toward the end of the Triassic, a series of geological events caused Pangaea to slowly begin to break into two. At the end of the Triassic Period there was a mass death, in which large land animals became extinct, but the dinosaurs survived, subsequently evolving into a variety of forms, becoming highly successful and surviving as total species for hundreds of millions of years.

During the Jurassic Period, the single land mass, Pangaea, split into two, forming Laurasia in the north and Gondwana in the south. Despite this separation, similarities in their fossil records show that there were some land connections between the two continents early in the Jurassic. These regions became more distinct later in the period.

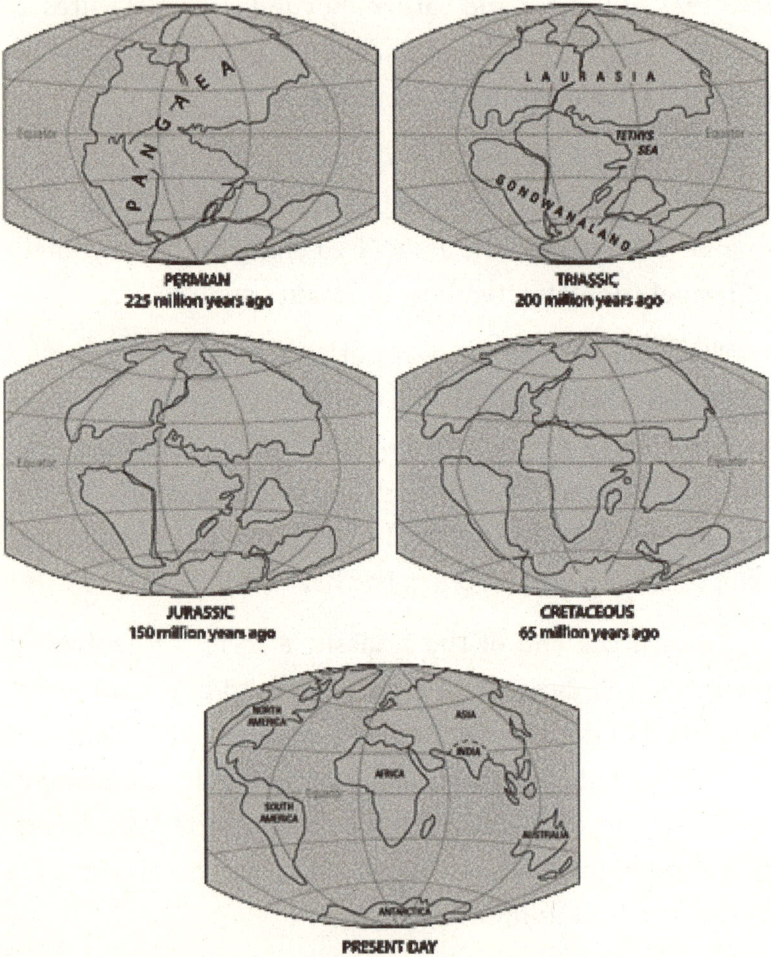

Evolution of the continents. Earthguide at Scripps Institution of Oceanography

Temperatures fell slightly, although it was still warmer than today due to higher amounts of carbon dioxide in the atmosphere. Rainfall increased when large seas appeared between the land masses, and this allowed plants to grow over huge areas. Some of this vegetation became the fossil fuels that we mine today. Elsewhere there were forests of tall conifer trees. The plentiful plant supply facilitated the evolution of plant-eating sauropods such as Apatosaurus, Diplodocus and Brachiosaurus. They were some of the largest animals to have ever walked Earth, and their herds dominated the landscape by the end of the Jurassic.

During the Cretaceous, the land began to further separate into the shape of the continents we recognize today. The dinosaurs were now evolving independently in different parts of the world, which increased their diversity. The first snakes evolved during this time, and by the end of the Cretaceous, flowering plants were a much more common part of Earth's ecosystem.

Various insect groups also appeared, including bees, which helped increase the spread of flowering plants. And mammals now included tree climbers, ground dwellers and even predators of small dinosaurs.

Sea levels rose and fell repeatedly during the Cretaceous Period. At the highest point there were many shallow seas separating parts of the continents we know today. At this time, Europe was made up of many smaller islands.

The fossil record shows that for the first 175 million years of their existence; dinosaurs took on many forms as they adapted to the changing environment and new species evolved.

But then 66 million years ago, over a relatively short time, dinosaurs disappeared completely. Many other animals, including pterosaurs, large marine reptiles, and ammonites, also died out in a mass extinction at the end of the Cretaceous Period.

It is not clear what lay behind this catastrophic event, but evidence points to an asteroid impact as the main culprit. Volcanic eruptions that caused large-scale climate change may also have been involved. The end of the age of the dinosaurs left gaps in ecosystems around the world. These were subsequently filled by mammals who went on to evolve rapidly.

During the Cretaceous extinction event, plants were less affected than animals because their seeds and pollen can survive harsh periods far longer. After the dinosaurs' extinction, flowering plants dominated Earth, continuing a process that had started in the beginning of this phase and continues to this day.

Initially the survivors were small, with birds the first to experience evolution to larger sizes. There were a couple of lineages of gigantic birds - predatory and herbivorous - but they weren't around for very long and went extinct.

There have been many close encounters of asteroids coming close to the Earth besides the major event that apparently led to the extinction of dinosaurs, and a similarly catastrophic asteroid encounter could recur in the future.

About every 450,000 years, Earth's magnetic poles flip in a phenomenon called geomagnetic reversal in which north pole becomes south, and south pole becomes north. There have been about 183 such reversals in the last 83 million years, but the last one was 780,000 years ago, and some believe that we're overdue for the next one.

Sometimes, the events are excursions rather than full reversals. That's when the field shifts for several hundred years and then returns to its original orientation, like the Laschamps event about 41,000 years ago. In an excursion, the field reverses in Earth's outer core while its inner core remains unchanged. These happen more frequently than full reversals, but their exact number and timing are more difficult to determine since their effects aren't global.

During Laschamps event, the Earth's magnetic field flipped for about 440 years. At its weakest, during the transition, the magnetic field dropped to around 5% of its current strength, and even when fully reversed, it was only about 25% of today's intensity. This significant weakening reduced the planet's magnetic shielding, allowing more cosmic rays and solar particles to penetrate the atmosphere. Evidence of this is preserved in geological records worldwide, like lava flows, sediment cores, and even ice cores. Another notable excursion is recorded in the sediments and volcanic rock Mono Lake in California that occurred 34,000 years ago.

Life on Earth

Life began around 3.5 to 4 billion years ago with simple single-celled organisms in Earth's primordial oceans. Over billions of years, these evolved into more complex forms through processes like photosynthesis and multicellularity. Fast forward to about 66 million years ago, after the extinction of dinosaurs, mammals began to diversify, and primates emerged around 55-60 million years ago. Early primates were small, tree-dwelling creatures with grasping hands and forward-facing eyes for depth perception.

The human lineage diverged from other primates much later. Around 25-30 million years ago, the superfamily Hominoidea (apes) split off, including ancestors of modern gibbons, orangutans, gorillas, chimpanzees, and humans. By 13-15 million years ago, the Hominidae family (great apes) emerged, and around 6-7 million years ago, the line leading to humans separated from that of chimpanzees, our closest living relatives based on genetic and fossil evidence.

Species like Australopithecus afarensis (e.g., the famous "Lucy" from 3.2 million years ago in Ethiopia) walked upright, a major adaptation for life on the savanna as forests receded due to climate shifts. Their brains were still small, about 400-500 cubic centimeters, like a chimpanzee's.

Homo habilis, dubbed the "handy man," appeared in East Africa and he used simple stone tools. The brain size grew to around 600-700 cc, reflecting increased cognitive ability.

Homo erectus (1.9 million to 110,000 years ago): This species spread across Africa, Asia, and Europe, mastering fire, refining tools, and showing early signs of social organization. Their brains ranged from 850-1,100 cc.

Species like Homo heidelbergensis and Homo neanderthalensis (Neanderthals) emerged, with larger brains (1,200-1,700 cc) and complex behaviors like hunting, art, and possibly language. This represented a cognitive revolution. Neanderthals, thriving in Europe and Asia until about 40,000 years ago, interbred with early modern humans, leaving traces in non-African DNA today.

According to the dominant theory, modern humans (Homo sapiens) migrated out of Africa around 60,000 to

100,000 years ago, spreading across the globe and gradually replacing or outcompeting other archaic human species, such as Neanderthals in Europe and Denisovans in Asia. Genetic and fossil evidence supports this model, showing that all modern human populations share a common African ancestry, with some interbreeding with archaic species along the way (e.g., about 1-2% Neanderthal DNA in non-African populations, and about half of that in African populations).

According to an alternative idea, called the multiregional hypothesis, modern humans evolved simultaneously in multiple regions (Africa, Europe, Asia) from different populations of archaic humans (like Homo erectus) that were already spread across the world. In this view, gene flow between these regions kept humans a single species, but regional differences developed over time.

Human ancestry is not quite a single stream; it is a braided flow.

Life beyond Earth

Space offers one solution to the many challenges of modern societies. We can begin dealing with our current problems through space colonization for survival of humanity beyond Earth, to expand living space, and to tap into extraterrestrial resources. The Moon and Mars are the natural next frontier due to their proximity, and permanent human colonies on them will provide valuable information on how to explore other places.

Firstly, let's consider the size and gravity at the Moon and Mars. The Moon's mean diameter is less than a third of that of Earth, and its gravitational force is only about one-sixth that of Earth's gravity. Mars has half the radius of Earth and a surface area that is 28.4%, and gravity that is about 38% that of Earth.

Conditions on the Moon and Mars are harsh. Challenges to settlement include the intense ionizing radiation that falls on the Martian surface, and the fine, toxic dust that covers the planet. Mars has an atmosphere, but it is unbreathable and thin. Surface temperatures fluctuate widely. Therefore, at first it will have to have habitats with highly trained humanoid robots. This will be followed by terrestrialized accommodations where insects, birds, and animals can be shielded from harmful radiation before it can be ready for long-term human residents.

Mars has an orbit that is more eccentric than Earth's, increasing temperature variations over the course of the Martian year. It has no rain and virtually no clouds, so although cold, it is permanently sunny excepting when it has dust storms that can last weeks. These dust storms would affect electricity production from solar panels for long periods and interfere with communications with Earth.

It has been proposed that a constellation of 81 low-orbit satellites around Mars would be good enough for planetwide coverage.[5] It would provide a local communications system that would be an extension of Earth's internet and be like Starlink, a constellation of satellites in low Earth orbit operated by SpaceX.

[5] Pfandzelter, T., Bermbach, D. 1st ACM MobiCom Workshop on Satellite Networking and Computing (SatCom '23), October 6 (2023), Madrid, Spain

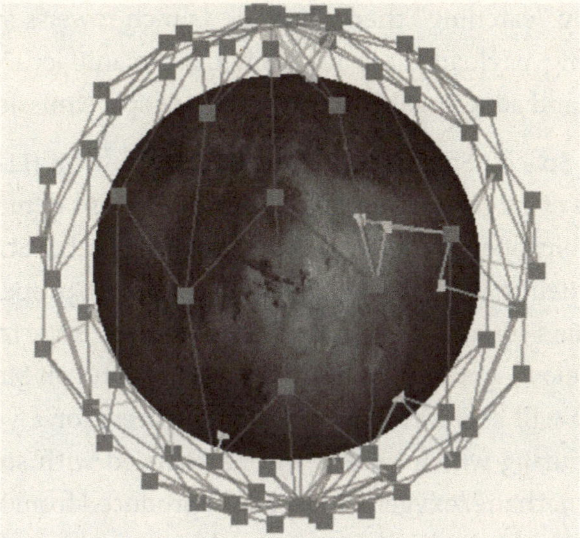

A satellite constellation around Mars with 81 satellites, with nine orbital planes of nine satellites each at 1,120km altitude.

Many space advocacy groups, such as the Mars Society and the National Space Society, and private organizations such as SpaceX, back the idea of colonization. Public space agencies (including NASA, ESA, Roscosmos, ISRO, the CNSA, among others) have explored colonization, but are primarily focused on further robotic exploration of Mars and the possibility of crewed landings.

There are several proposals for missions to Mars. SpaceX's Starship has been pivotal, with its development aiming at making life multiplanetary, particularly targeting Mars. Starship is a two-stage fully reusable super heavy-lift launch vehicle that is the most massive and most powerful vehicle ever to fly. It was developed with the intention of lowering launch costs using economies of scale, by reusing both rocket stages by "catching" them with the launch tower's systems, increasing payload mass to orbit, mass-manufacturing the rockets and adapting it to a wide range of space missions.

The SpaceX Starship will fly to low Earth orbit (LEO) with 100 tons of cargo and then refuel it with 600 tons of methane/oxygen bipropellant delivered to orbit by six tanker Starships. This refueling would provide it with sufficient propellant to fly to Mars on a six-month Conjunction-class trajectory (see figure below), get into Mars orbit, and then land on Mars. The Starship will serve as the home for the crew for a year and a half, during which time it will be refueled with some 600 tons of methane/oxygen bipropellant produced from Martian carbon dioxide and water. This would be enough to fly back to Earth on a six-month trajectory carrying the crew and ten tons of cargo.

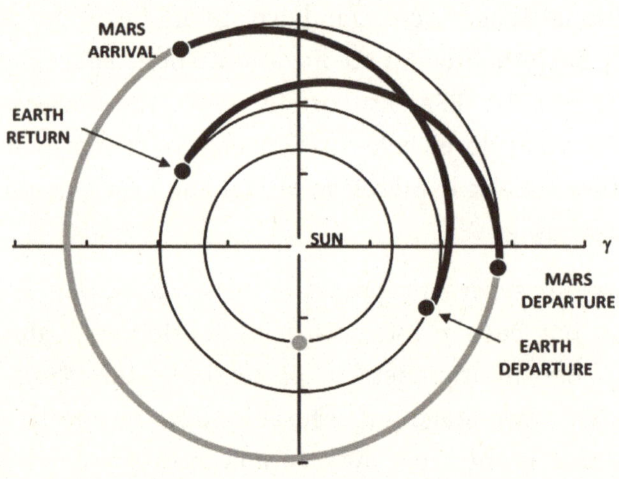

Conjunction: Roughly equal-length trips to Mars and back, with a long stay (up to 1.5 years)

Conjunction missions involve a 2.5-year roundtrip to Mars, with roughly equal outbound and inbound legs of 6 to 8 months, and up to 1.5 years on the Martian surface.

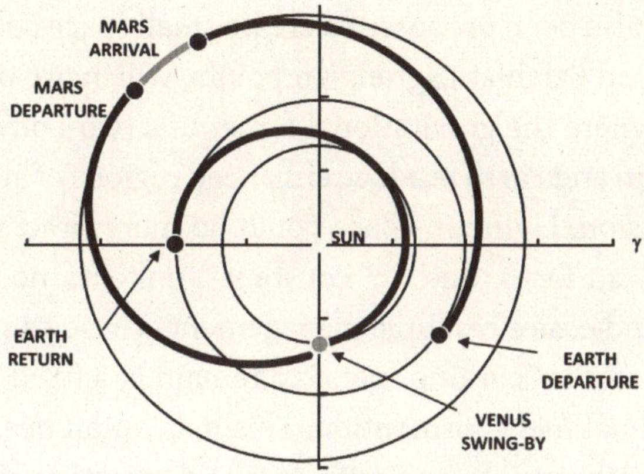

Opposition: One short trip, one very long trip, and a very brief stay on Mars (30 days)

Opposition missions will have a roundtrip time of 2 years, but almost all that time would be spent in transit on two very unequal transfer legs and only about 30 days spent on or orbiting Mars.

In a modified proposal, a smaller spaceship called Starboat will be used to deal with the complication of Starship being too heavy to take off from Martian surface.[6] Starboat would be like Starship but scaled down by a factor of five in mass. "It could do a direct return from the Mars surface to Earth using 120 tons of propellant or perform a low-Mars-orbit rendezvous using just 50 tons of propellent, with a single tanker in low Mars orbit being able to support five such

6 Zubrin, R. The Mars Dream Is Back — Here's How to Make It Actually Happen. The New Atlantis (2025)

return flights. It could also be lifted to Earth orbit fully fueled by a single Starship and sent directly to Mars with five tons of cargo without any Earth-orbit refueling, or 25 tons of cargo with a single tanker refueling."

It has also been proposed to create small space colonies in orbit around Earth at Lagrangian points, which are positions in space where the gravitational forces of a two-body system like the Sun and Earth produce enhanced regions of attraction and repulsion. Living in space could be more practical than planetary surfaces due to constant sunlight, no seismic activity, and easier resource management. These can be used by spacecraft as "parking spots" to remain in a fixed position with minimal fuel consumption. This may entail health risks related to low gravity environments and radiation exposure. Creation of gravity like that on Earth, perhaps by using rotation of spacecrafts, will likely be a high priority of future inhabitants of space.

Colonies away from the comfortable Earth-like environments would need to be largely self-sufficient, recycling water, air, and growing food. This requires breakthroughs in closed-loop life support systems and possibly in-situ resource utilization for materials and energy. This is essential if life in future colonies is to spread beyond the solar system. One can visualize three stages for the colonies:

Pre-settlement: small outpost (near-term future)
In-settlement: permanent settlement (medium-term future)
Post-settlement: self-sufficient society (long-term future)

Initial colonies might be economically driven by tourism, mining, or scientific research. Over time, these would become self-sustaining economies, based on industries like asteroid mining, space manufacturing, and scientific research. These colonies may end up developing unique cultures. As humans spread beyond Earth, other life forms will spread as well. Evolution will take a different dynamic as in the future humans will be the custodians of all life on Earth and will be the main driver of collective existence.[7]

It is hoped that this situation will be achieved by lunar bases or permanent human presence on Mars by the mid-21st century. The long-term vision is to create a multi-planetary civilization, with humans potentially spreading across the solar system, utilizing resources from asteroids and moons of the planets, perhaps even beyond the Solar system to other nearby star systems and in the distant future, beyond the Milky Way Galaxy.

But how will animal life be seeded in these remote places? Once the host place has been prepared to have the right environment of plant life and stocked with food, the first step will be to send eggs by spaceship to create an oviparous life which will over time comprise a rich ecosystem.

Next is the issue of transporting humans. One idea is to send a robotic mission to a habitable planet with frozen early-stage human embryos or the technological or biological means to create human embryos. Modern medicine has made it possible to store frozen embryos in various low-development stages. Upon arrival of the embryo-carrying spacecraft at the target planet, fully autonomous robots would build the first settlement on the planet and

---

[7] Futuyma, D.J., Kirkpatrick, M. Evolution. Sinauer Associates (2017)

start growing food. Thereafter the first embryos could be unfrozen or created using biosequenced or natural sperm and egg cells and raised by robots.

Another approach is to take advantage of the relativistic effect that clocks on spaceships moving close to the speed of light experience very little time elapse. Thus, a spaceship going at 0.99 times the speed of light will experience a passage of time that is only 14 percent of that for earthbound observers. The nearest star to Earth, after our own Sun, is Proxima Centauri, located 4.37 light-years away in the southern constellation of Centaurus. For the crew on the spaceship going at 0.99 times the speed of light, only about six months would have passed. At speeds virtually the same as the speed of light, the crew would feel that they arrived literally instantaneously. The challenge would be to find a physical process to accelerate the spaceship to this high speed and then to slow it down upon arrival at the destination without risking the lives in the spaceship due to high acceleration and de-acceleration.

One aspect of stellar existence is the great distances between stars, dwarfing the distances between planets in our solar system. Distances beyond a few light years make interstellar travel nearly an impossible undertaking.

The future of space colonies hinges on technological breakthroughs and overcoming significant biological challenges. Reality will likely unfold through incremental steps, each building on the successes and failures of the previous. Our progeny will evolve with all life on Earth, while influencing and being influenced by technology.

Landers and rovers have successfully explored the surface of Mars and delivered information about conditions on the ground. The first successful lander, Viking 1, touched down on the planet in 1976. Crewed missions to Mars have been proposed, but no person has set foot on the planet, and there have been no return missions.

Data transfer between Mars and Earth is handled by the Mars Relay Network relaying the data first onto computers on the orbiters deployed around the planet, which then forward it to Earth ground stations (NASA's Deep Space Network). The network includes NASA's Mars Reconnaissance Orbiter, Mars Odyssey, and MAVEN, and two European Space Agency missions, Trace Gas Orbiter and Mars Express. They capture data from NASA's Perseverance and Curiosity rover.

This reduces the required mass for Mars landers, as they only need power and antenna for radio communication to Mars orbit rather than direct-to-Earth communication. Since the several-minute round-trip time for signals between Mars and Earth is too large to run all these systems on Earth computers, Mars will require its own "Mars cloud".

Innovations like solar sails and new ion thrusters are being explored for their potential in reducing travel costs and time, offering a more sustainable approach to propulsion beyond traditional chemical rockets. Discussions around interplanetary travel also touch on destinations like Venus, which is very hot although it is like the Earth in mass, Jupiter's Galilean moons, and Saturn's Titan.

Mystery of consciousness

The efforts to intensify exploration of space have a complementary philosophical side. To understand the past and the unfolding future we must recognize that all our knowledge resides in our consciousness. We take this fact for granted but don't understand it scientifically and throw up hands and just say that it is simply a collective associated with our minds. By mind, we normally mean the entity where memories are stored, and how intelligence is used to make judgements, and both capacities exist in computers as well.

Exploring this matter further, we arrive at a scientific distinction between mind and consciousness.[8] Mind does the manipulation of information whereas consciousness is the *awareness* that exists at the back of the manipulation which computers lack. According to the Orthodox Copenhagen Interpretation of quantum theory, which is the dominant way to understand the subject, consciousness is complementary to materiality,[9] and it is not a property of matter. Consistent with this, the recently published One Observer Theorem[10] implies that consciousness transcends materiality.

Given these results, one must take consciousness to be the non-changing ground of reality that serves as a kind of light that illuminates the contents of the mind. The structures of the brain constituting the mind are essential for the experience

[8] Kafatos, M. and Nadeau, R. The Conscious Universe: Parts and Wholes in Physical Reality, Springer (2000)

[9] Kafatos, M.C. Science, Reality & Everyday Life. Prisma Press, Auroville, India (2024); Kafatos, M.C., Banerji, D., and Struppa, D. (editors). Quantum and Consciousness Revisited. DK Publisher (2024)

[10] Kak, S., No-go theorems on machine consciousness. Journal of Artificial Intelligence and Consciousness 10, 237-247 (2023)

of consciousness and their evolution in time across different species is of abiding relevance. We believe that all animals experience consciousness although they may do it in a variety of ways. Some of the structures required for consciousness in humans are in the brainstem and diencephalon that operate in conjunction with neocortical connectivity and recurrent processing.

If consciousness is not a product of materiality, then there is likely a connection between the outer cosmos and the inner cosmos of human awareness to which we have access through the mind. It is likely that such a connection makes it possible for the human imagination to anticipate universal truths. If one were to consider ancient philosophers and sages from across the world, it is this process by which they arrived at judgments and assertions about the nature of reality that appear true even now.

A quote on evolution on earth from the nearly two-thousand-year-old text Yoga Vāsiṣṭha (6.1.21) that anticipates early Earth without life and then gradual evolution with H. Sapiens appearing at the end of a chain:[11]

> I remember that once upon a time there was nothing on this Earth, neither trees and plants, nor even mountains. For a period of eleven thousand years (4 million Earth years) the Earth was under lava. In those days there was neither day nor night below the polar region: for in the rest of the earth neither the sun nor the moon shone. Only one half of the polar region was illumined. Then asuras ruled the earth. They were deluded, powerful and prosperous, and the earth was their playground. Apart

[11] Venkatesananda, S. Vasistha's Yoga. SUNY Press (2010)

> from the polar region the rest of the earth was covered with water. And then for a very long time the whole earth was covered with forests, except the polar region. Then there arose great mountains, but without any human inhabitants. For a period of ten thousand years the earth was covered with the corpses of the asuras.

In a similar fashion, several descriptions in the Puranic texts of India anticipate the end of Earth:[12] "At first there is a drought that extends for many years, and men and creatures perish by millions. Then seven blazing suns appear on the horizon, sucking up all the water from the rivers and seas, and all is reduced to ashes by the heat. Then there arise in the sky large masses of clouds, monstrous in appearance, with terrible lightning. They then pour rain over the earth. The rain soon floods the entire surface of the earth. When the entire world is thus flooded, no movable or immovable creatures or beings can remain alive, and all are destroyed. This water is later transformed into fire, and still later into a wind."

Next we present anomalous events as evidence of the workings of consciousness in the large, although one could also explain them as coincidences. The difficulty with the latter explanation is that the probability of choosing numbers mentioned here is extremely small. These cases are different from speculations in ancient texts (somewhat like science fiction) describing airplanes, space travel, embryo transplantation, weapons that can destroy the world and so on that have come to pass because they are general ideas which do not involve specific numbers. The figure of 8.64 billion

---

[12] Dimmitt, C., van Buitenen, J.A.B. Classical Hindu Mythology: A Reader in the Sanskrit Puranas. Temple University Press (2012)

years for the universe cycle in Indian cosmology, which is not too different from the currently accepted value of the time since the Big Bang, is widely known. Here we speak of two other numbers that are not well known.

Speed of Light. The mention of the 14th century knowledge of the speed of light in India is an example of spontaneous discovery in the manner discussed by Hadamard in his book on how scientific creativity occurs.[13] The speed is found in the well-known commentary on the Rigveda by the medieval scholar Sāyaṇa (1315-1387), prime minister in the court of the Vijayanagar Empire. It associates the speed of 2,202 yojanas in half a nimesha with the sun (or sunlight). The time measure of nimesha is well attested in Indian astronomical and encyclopedic texts as being equal to 16/75 seconds.

The yojana is variously defined by different writers. It was defined in two different ways by leading astronomers of the time. Likewise, it seems to have been used in two different ways for terrestrial applications.

According to the Arthaśāstra, the famed third century BCE book on polity and economics, a yojana is 8,000 dhanus (bow). For a long time, Western scholars took a dhanus to be equal to six feet or 1.83 meters,[14] and with this Sāyaṇa's number is 3.02×10^8 meters per second which corresponds closely to the correct value of the speed of light.

There is no way this speed could have been measured in medieval India. Until just over 200 years ago it was not even

[13] Hadamard, J., The Psychology of Invention in the Mathematical Field, Dover, New York (1954)

[14] Basham, A.L. The Wonder That Was India, Sidgwick and Jackson (2000)

known in the Western tradition that light had finite speed. In 1676, Rømer calculated this speed in terms of the speed of earth's rotation around the sun, and his value was about 26% less than the modern value. Sāyaṇa could not have obtained this figure from the West or anywhere else and, therefore, it can only be seen as an inspired guess.

Distance to the sun. Until modern times, the solar system was taken to be much smaller than its currently accepted value. Pappus of Alexandria writing in the 4th century took the distance to the sun to be about 490 earth radii. At the beginning of modern Western astronomy, Tycho Brahe took the distance between the earth and the sun to be 8×10^6 kms and Kepler took it to be about 24×10^6 kms.

In the Indian Purāṇas, the earliest of which predates fifth century, the Sun is associated with two "axles": the first one is 15.7×10^6 yojanas long and the second one is 45,500 yojanas long[15] (section 2.8 of the Viṣṇu Purāṇa). One can presume that the longer axle represents the distance to the earth and the shorter axle represents the deviation from the circular path. Speaking of this astronomical reference, we cannot prove that it is from the earliest layer of the text, but there is general scholarly consensus that the text is at least 1,500 years old.

One of the variant definitions of yojana, which was prevalent in Kashmir (a region where many Purāṇas were written), is given by Stein[16] (footnote to verse 393) where he takes it to be about 6 miles (Stein: "The direct distance

[15] Wilson, H.H. (tr.), The Vishnu Purana, (1840)

[16] Stein, M.A. (tr.), Kalhana's Rajatarangini. A Chronicle of the Kings of Kasmir, Vol. 1. Delhi: Motilal Banarsidass (1961). Reprint of the 1900 edition.

by road from Srinagar to Vijebror is about thirty miles. This corresponds exactly to five yojanas."). If this yojana was meant (6 miles or 9.6 km), then the distance to the sun is about 150.72 × 10^6 kms. Given that the distance between the two cities is an approximate value, the distance to the sun can be taken to be identical to the modern value of 149.598 × 10^6 kms.

One might complain that we are picking a value of yojana that was only used regionally. This criticism is valid, but we are not trying to find a scientific basis to the choice of the distance to the sun. Rather, we are showing that the choices presented by intuition, without any firm basis in physical theory, can be surprisingly accurate. Accepting the validity of such a way of knowledge is against the mainstream program of science and current schoolbook theories of mind.

It is indeed possible that even before modern telescopic astronomy, which did not exist in India, a reasonable estimate of the distance to the sun could be made based on durations of lunar and solar eclipses. But there is no evidence that such calculations were made by astronomers.

There are other anomalous precise numbers in the ancient texts that include the size of the earth in the accounts of Eratosthenes, Āryabhaṭa, and al-Bīrūnī. The apocryphal account of Eratosthenes's measurement of the size indirectly by measuring the shift in the shadow of the sun at noon between Syene and Alexandria is not credible. The story is that Eratosthenes had heard from travelers that the bottom of a well in Syene (now Aswan, Egypt) was illuminated at noon on the summer solstice, indicating that the sun was directly overhead. Eratosthenes then measured the angle of a shadow cast by a stick at noon on the summer solstice in Alexandria, taken to be at 5,000 stades from Syene, and found it made an angle of about 7.2 degrees, or about 1/50 of a complete circle. From this he calculated that the diameter of Earth was 250,000 stades. But Goldstein convincingly shows that no precise measurements were involved, and the correctness of the result should be seen as a coincidence.[17]

In the traditional explanation of the workings of mind, habits picked up in childhood and in school are the impediments that prevent one from being connected to one's intuition. Real creativity requires challenging dogma as well as one's own certitudes. One sees unexpected connections, which is an element of creativity, in extraordinary states of mind. Looking within can reveal unexpected knowledge about the universe for we are a part of the universe.

There is also the fascinating coincidence of perfection in the Sun-Moon proportions known to the sages. The Moon's nearly exact distance from Earth is 108 times its diameter, and the Sun's distance from Earth is also 108 times its diameter.[18] The number 108 holds special significance in many cultures and natural patterns. The diameter of the Sun is also about 108 times the diameter of Earth.

The Sun and Moon work together, bringing light to Earth—one during the day and the other at night. The Moon acts like a radiant mirror, reflecting sunlight, and its movement has fascinated scientists for centuries. In ancient

[17] Goldstein, B.R. Eratosthenes on the measurement of the Earth. Historia Mathematica 11, 411-416 (1984)

[18] Kak, S. The structure of the Ṛgveda. Indian Journal of History of Science 28, 71-79 (1993)

India, Vedic texts called the Moon "Soma," a symbol of the mind, time, and nature's nourishing power.

The Moon's gravity helps keep Earth's tilt stable, which ensures we have relatively steady climates. It also controls ocean tides, which are critical for marine life. For example, fish and coral, like those in the Great Barrier Reef, time their reproduction based on the Moon's cycle. During full and new Moons, they release eggs and sperm at the same time to maximize fertilization. Birds and other animals also rely on the Moon, with seabirds laying eggs when the full Moon provides extra light. Migratory birds use the Moon to navigate during their long journeys at night.

The Moon's phases might even influence human moods. Ancient traditions suggest lunar cycles affect hormones, and some modern theories support this. Farmers in many cultures used the Moon to decide when to plant and harvest crops. People also fasted during specific lunar phases, as it was believed to help with bodily cleansing. The Moon affects soil moisture and plant growth at night, showing how deeply connected it is to life on Earth. Yet, despite all these benefits, current plans aim to exploit the Moon's resources rather than appreciate its significance.

The stars in the sky were pictured as belonging to the figure of a cosmic man. This metaphor represents relationships in the universe across scales. It appears that the actual connection between stars and living beings was based on the identity between basic biological rhythms and astronomical periods.

Fiddler crabs, in their natural habitat on the shore, burrow themselves during high tide, emerging when the tide recedes to feed, mate, and challenge each other. When these crabs are removed to the laboratory and held in an incubator with constant conditions, they still run around in their containers during the time of each low tide. Not only crabs but all living creatures have extremely precise inner cycles.

In humans the menstrual period has by tradition been taken to correspond to the moon's motion and "menses" means lunar month. New research supports this:

> In a study of a number of women with variable onset of menstrual periods, artificial illumination of the bedroom through the 14th to 17th nights following the onset of menstruation resulted in the regularization of the period, with the period length coming very close to 29.5 days, the natural synodic month. That this period is a biologically significant one for the human species is further suggested by the fact that the average duration of pregnancy (from ovulation to birth) in the human is rather precisely nine 29.53 synodic months. *Encyclopaedia Britannica* (1994; Macropaedia article on Animal Behaviour, p. 761)

One should note the distinction between lunar and free running circalunar cycles. A lunar cycle is in step with the motions of the moon. The menstrual cycle is a free running cycle with the same period as that of the moon. One might assume that entrainment to the lunar cycle was triggered by moonlight. In the living under artificial lights of modern times it is easy to see how the direct correlation with the Moon's motion has been lost.

It is a surprise that the fundamental circadian rhythm inside us is not the 24-hour one related to the motion of the Sun but rather the 24 hour 50 minute one according to the period of the Moon, since each moonrise is 50 minutes later than the preceding one. We share this approximately 24-hour-50-minute clock with monkeys and other non-humans.

This 24-hour-50-minute clock was discovered by the moderns only about 50 years ago in experiments on a blind squirrel monkey. The activity of this monkey was recorded night and day for a period of three years, and it was discovered that her rhythms drifted later each day by an average of about 46 minutes. We do not know if the deficit of four minutes from the moonrise period is due to the reference with respect to the stars. The monkey kept her own time, unaffected by the activities around her.

That this connection might have been known in the ancient world is suggested by the fact that the Moon (Soma) is called the "lord of speech" in the Rigveda (RV 9.26.4; 9.101.5). It is taken to awaken eager thoughts (RV 6.47.3). Many references connect the Moon with the mind as in RV 10.90 where it is stated that the mind is born of the Moon. This is expressed in the Śatapatha Brāhmaṇa 8.1.2.7 as the slogan: "the mind is the Moon."

It is reasonable to assume that with their emphasis on time bound rituals and the calendar, the Vedic Indians of the 3rd millennium BCE knew many biological periods. Having seen rhythms matched closely to the principal astronomical periods, they must have further assumed that there were less obvious cycles that were matched to the motions of the other heavenly bodies. This must have been a motivation in determination of the planetary periods.

The gestation periods for mammals also have intriguing relationships to astronomical numbers. Perhaps such connections were the basis for marking out certain animals as special icons. The averages of some of the gestation periods are:

ass	365 days
sacred baboon	183 days
cow	284 days
dog	61 days

It is no wonder then that the ass is used as a symbol of the year in Śatapatha Brāhmaṇa, an ancient Indian text from a period somewhat after the early Vedic age. The sacred baboon's gestation period of exactly half the solar year is likely to have played a role in the special significance attached to it by the Egyptians. That the gestation periods for the human and the cow are about the same may be another reason for the sacredness assigned to the cow. In other words, knowledge of biological periods was significant in the choice of sacred symbols.

The importance that the Moon has on the inner biological clock of the humans likely led to the adoption of the luni-solar calendar in India. Its use can be seen at least as early as the third millennium BCE.

Surprising new Findings about the early universe

In the standard model of cosmology, the universe was extremely hot and dense in the first moments after the Big Bang. As the universe cooled, conditions became just right to give rise to the building blocks of matter – the quarks and electrons of which we are all made. A few millionths of a second later, quarks aggregated to produce protons and neutrons. Within minutes, these protons and neutrons combined into nuclei. As the universe continued to expand and cool, things began to evolve more slowly.

It took 380,000 years for electrons to be trapped in orbits around nuclei, forming the first atoms. These were mainly helium and hydrogen, which are still by far the most abundant elements in the universe. Gas and dust clouds collapsed under their own gravity, forming stars. Small clumps of gas and dark matter merged to form protogalaxies, which then merged further. Present observations suggest that the first stars formed from clouds of gas around 150–200 million years after the Big Bang. Heavier atoms such as carbon, oxygen and iron, have since been continuously produced in the hearts of stars and spread throughout the universe in spectacular stellar explosions called supernovae.

But stars and galaxies do not tell the entire story. Astronomical and physical calculations related to anomalous motions of stars in galaxies suggest that the visible universe is only a tiny amount (4%) of what the universe is made of. A very large fraction of the universe, in fact 26%, is made of an unknown type of matter called "dark matter". Unlike stars and galaxies, dark matter does not emit any light or electromagnetic radiation of any kind, so that we can detect it only through its gravitational effects.

An even more mysterious form of energy called "dark energy" accounts for about 70% of the mass-energy content of the universe. This idea stems from the observation that all galaxies seem to be receding from each other at an accelerating pace, implying that some invisible extra energy is at work. But there is no evidence for the postulated dark matter and dark energy that are supposedly 96% of the universe. Furthermore, the result of the continuing accelerating expansion of the universe will be the death of the universe and all life everywhere, which implies that the universe has had no meaning whatsoever.

Ours is thus a time of a crisis in cosmology. Astronomers previously thought it took many billions of years for galaxies to become stable enough to develop so-called "bars," ribbons of stars and gas that cut across the core of a galaxy. Bars form in spiral galaxies when the orbits of stars near the galaxy's nucleus become erratic and stretched out, and they grow as gravity collects more nearby stars. Eventually, a large percentage of the stars within the galaxy's heart get caught up in the bar. When astronomers detect a star bar, they usually interpret it to mean the galaxy has reached a settled phase.[19]

It was not that long ago that based on a study of 2,000 galaxies from a Hubble Space Telescope census NASA proclaimed that barred spiral galaxies were "latecomers" to the universe. That sample found far fewer Milky Way-like galaxies 7 billion years ago, when the universe was about

---

[19] Hawking, S. A Brief History of Time. Bantam (2011)

half its current age. Only 20 percent of the spiral galaxies in the distant past possessed bars, compared to nearly 70 percent of their more-modern cohort. Those observations led astronomers to conclude that even massive galaxies closer to the dawn of time would have been clumpier, with ill-defined structures.

Photograph of Horse head nebula/ Bryan Goff (Unsplash)

And now the most distant and thus earliest rotating disk galaxy named REBELS-25 challenges these theories of galactic evolution. Discovered by a team of astronomers using the Atacama Large Millimeter/submillimeter Array (ALMA), the galaxy shows hints of spiral arms like the highly evolved "modern" galaxies like our own Milky Way. The conclusion was that dynamically settled disc galaxies can form as early as 700 Myr after the Big Bang.[20]

---

[20] Rowland, Lucie E.; et al. REBELS-25: Discovery of a dynamically cold disc galaxy at z = 7.31. Monthly Notices of the Royal Astronomical Society 535 (3), 2068–2091 (2024)

The high fraction of spiral galaxies at high redshifts supports the scenario that many spiral galaxies formed at least 12 billion years ago and settled to dynamically cold, thin disk spirals like the Milky Way around a billion years later. For context, very few spiral galaxies were known to be older than 8 billion years before we had the powerful sight of JWST.

CEERS-2112 is the most distant barred spiral galaxy observed as of 2023.[21]The light observed from the galaxy was emitted when the universe was only 2.1 billion years old. It was determined to be similar in mass to the Milky Way. The galaxy has a stellar mass of 3.9 billion times that of the Sun, comparable with that of the Milky Way 11.7 billion years ago.

Understanding when and how spiral galaxies formed helps to answer big questions, like how the universe evolved as a whole, and still-big-but–slightly-smaller questions, like piecing together the mechanisms that formed spiral arms. Realizing spiral galaxies are older than we thought brings us one step closer to understanding spiral galaxy evolution.

REBELS-25 appears to have characteristics like those of the roughly 13.6-billion-year-old Milky Way. These include an elongated "central bar" of stars with hints existing that it may even possess spiral arms. This would lead to REBELS-25 breaking another record as the most distant and earliest spiral galaxy ever seen. The current record holder is CEERS-2112, seen by the James Webb Space Telescope

---

[21] Costantin, L. et al. A Milky Way-like barred spiral galaxy at a redshift of 3. Nature. 623 (7987), 499–501 (2023)

(JWST) when the universe was around 2.1 billion years old. The discovery is not only a record-breaker but another piece of mounting evidence that suggests cosmologists need to revise their previous notions about galaxy evolution.

Barred spiral galaxy UGC 6093 is an active galaxy with a central supermassive black hole, heating up and causing the galaxy's core to shine brightly. Credit: ESA/Hubble

A recent study examined images of galaxies and calculated how the fraction of spiral to non-spiral galaxies changed over time. The team studied light that had traveled as far as 12 billion years (out to a redshift of 4). They were surprised to find that these early galaxies were not only more common than previously imagined, but they had already developed fully formed spiral arms and disks not unlike what we find in our own local group of galaxies.

If the rotation of galaxies averages about 200-300 km/s, and a full rotation takes say 250 million years, then a galaxy that is 1 billion years old has already rotated four times. Is this sufficient time to form spiral arms? If galaxies were smaller after the time of the Big Bang, then the full rotation time would be much shorter in time.

Important questions arise: how are the shapes of galaxies, their stellar contents, whether they are members of clusters of galaxies or isolated field galaxies, as well as stellar evolution into advances stages, how the end stages of stellar remnants are perhaps tied into next generations of stars and planetary systems which would be directly tied to life in the universe?

The story of the universe is perhaps much more diverse than we understand now, making it possible for life to develop and evolve beyond what occurred in the Solar system in the last five billion years.

As we ponder how we got where we are now in time, one thing is clear, that the future may be much richer than anything we can possibly imagine.

The origin of the universe and life

In the late 1920s, the American astronomer Edwin Hubble made the significant discovery that distant stars and galaxies are receding from Earth in every direction. The velocities of recession increase in proportion with distance which implies that the universe is expanding.

Hubble's hypothesis of an expanding universe means that the universe was denser at a previous time. Indeed, one can assert that in the beginning the universe was a very small and infinitely hot mass. A huge explosion, known as the Big Bang, then sent matter and energy stretching in all directions.

This Big Bang hypothesis leads to testable deductions. One such conclusion is that the temperature in deep space today should be several degrees above absolute zero. Observations showed this to be correct. In fact, the Cosmic Microwave Background Explorer (COBE) satellite launched in 1991 confirmed that the background radiation field has exactly the spectrum predicted by a Big Bang origin for the universe.

As the universe expanded, matter collected into clouds that began to condense and rotate, forming the forerunners of galaxies. Within galaxies, including our own Milky Way galaxy, changes in pressure caused gas and dust to form distinct clouds. In some of these clouds, where there was sufficient mass and the right forces, gravitational attraction caused the cloud to collapse. If the mass of material in the cloud was sufficiently compressed, nuclear reactions began, and a star was born.

Some proportion of stars, including our sun, formed in the middle of a flattened spinning disk of material. In the case of our sun, the gas and dust within this disk collided and aggregated into small grains, and the grains formed into larger bodies called planetesimals ("very small planets"), some of which reached diameters of several hundred kilometers. In successive stages these planetesimals coalesced into the nine planets and their numerous satellites. The rocky planets, including Earth, were near the Sun, and the gaseous planets were in more distant orbits.

The ages of the universe, our galaxy, the solar system, and Earth can be estimated using modem scientific methods. The age of the universe can be derived from the observed relationship between the velocities of and the distances separating the galaxies. The velocities of distant galaxies can be measured very accurately, but the measurement of distances is more uncertain. Over the past few decades, measurements of the Hubble expansion have led to estimated ages for the universe of between 7 billion and 20 billion years, with the most recent and best measurements within the range of 10 billion to 15 billion years.

The age of the Milky Way galaxy has been calculated in two ways. One involves studying the observed stages of evolution of different-sized stars in globular clusters. Such clusters occur in a faint halo surrounding the center of the Galaxy, with each cluster containing from a hundred thousand to a million stars. The very low amounts of elements heavier than hydrogen and helium in these stars indicate that they must have formed early in the history of the Galaxy, before large amounts of heavy elements were created inside the initial generations of stars and later distributed into the interstellar medium through supernova explosions (the Big Bang itself created primarily hydrogen and helium atoms). Estimates of the ages of the stars in globular clusters fall within the range of 11 billion to 16 billion years.

A second method for estimating the age of our galaxy is based on the present abundances of several long-lived radioactive elements in the solar system. Their abundances are set by their rates of production and distribution through exploding supernovas. According to these calculations, the age of our galaxy is between 9 billion and 16 billion years. Thus, both ways of estimating the age of the Milky Way galaxy agree with each other, and they also are consistent with the independently derived estimate for the age of the universe.

Radioactive elements occurring naturally in rocks and minerals also provide a means of estimating the age of the solar system and Earth. Several of these elements decay with half-lives between 700 million and more than 100 billion years (the half-life of an element is the time it takes for half of the element to decay radioactively into another element). Using these timekeepers, it is calculated that meteorites, which are fragments of asteroids, formed between 4.53 billion and 4.58 billion years ago. The same radioactive timekeepers applied to the three oldest lunar samples returned to Earth by the Apollo astronauts yield ages between 4.4 billion and 4.5 billion years, providing minimum estimates for the time since the formation of the moon.

The oldest known rocks on Earth are as old as 4 billion years. In Western Australia, zircon crystals encased within younger rocks have ages as old as 4.3 billion years, making these tiny crystals the oldest materials so far found on Earth.

The best estimates of Earth's age are obtained by calculating the time required for development of the observed lead isotopes in Earth's oldest lead ores. These estimates yield 4.54 billion years as the age of Earth and of meteorites, and hence of the solar system.

The origins of life cannot be dated as precisely, but there is evidence that bacteria-like organisms lived on

Earth 3.5 billion years ago, and they may have existed even earlier, when the first solid crust formed, almost 4 billion years ago. These early organisms must have been simpler than the organisms living today. Furthermore, before the earliest organisms there must have been structures that one would not call "alive" but that are now components of living things. Today, all living organisms store and transmit hereditary information using two kinds of molecules: DNA and RNA. Each of these molecules is in turn composed of four kinds of subunits known as nucleotides. The sequences of nucleotides in particular lengths of DNA or RNA, known as genes, direct the construction of molecules known as proteins, which in turn catalyze biochemical reactions, provide structural components for organisms, and perform many of the other functions on which life depends. Proteins consist of chains of subunits known as amino acids. The sequence of nucleotides in DNA and RNA therefore determines the sequence of amino acids in proteins; this is a central mechanism in all biology.

Experiments conducted under conditions intended to resemble those present on primitive Earth have resulted in the production of some of the chemical components of proteins, DNA, and RNA. Some of these molecules also have been detected in meteorites from outer space and in interstellar space by astronomers using radio-telescopes. Scientists have concluded that the "building blocks of life" could have been available early in Earth's history.

An important new research avenue has opened with the discovery that certain molecules made of RNA, called ribozymes, can act as catalysts in modem cells. It previously had been thought that only proteins could serve as the catalysts for specific biochemical functions. Thus, in the early prebiotic world, RNA molecules could have replicated themselves well before there were any protein catalysts.

Laboratory experiments demonstrate that replicating autocatalytic RNA molecules undergo spontaneous changes and that the variants of RNA molecules with the greatest autocatalytic activity come to prevail in their environments. Some scientists favor the hypothesis that there was an early "RNA world," and they are testing models that lead from RNA to the synthesis of simple DNA and protein molecules.

NASA's OSIRIS-REx spacecraft grabbed samples from the surface of the asteroid Bennu in 2020. The analysis of these samples shows that even in dark, cold regions of the early solar system, astonishingly diverse chemical processes gave rise to numerous building blocks of life.[22] Bennu's parent body apparently contained plenty of water, and some of the water evaporated and left behind a salty brine. This brine contained thousands of organic compounds, including 14 of the 20 amino acids found in terrestrial organisms, as well as all the nucleotide bases that make up our DNA and RNA. This means that the basic molecules of life existed in our solar system practically from the start.

---

[22] McCoy, T.J., Russell, S.S., Zega, T.J. et al. An evaporite sequence from ancient brine recorded in Bennu samples. Nature 637, 1072–1077 (2025).

Observations from the Kepler Mission suggest that about 2% of all Sun-like stars are expected to have Earth-size planets within the habitable zones of their stars. This means there could be as many as two billion Earth-sized planets in the Milky Way galaxy alone, and if all galaxies have number of such planets similar to the Milky Way, in the 50 billion galaxies in the observable universe, there may be as many as a hundred quintillion Earth-like planets.

Therefore, to contact life across the universe is an imperative for humans. It is conceivable that future advances in science will allow humans to travel to corners of the universe that are out of reach now.

2

Rockets and Spaceships

Humans have always dreamed about flying beyond Earth to the Moon and the stars. But it is only in the last half-century that rockets have become powerful enough to allow artificial satellites and humans to reach space.

Let us begin with the mythology associated with the Sun and the planets. This mythology is richest within the Indian tradition and so we will describe it first. Surya is the Vedic name and main deity associated with the Sun; it is the fire of the heavens and often associated with Vishnu as the preserver deity that sustains life. It is also Shiva as the light of consciousness.

The Moon is the consort of Shiva for it reflects light from the Sun and in the inner world it is the mind that is illuminated by the light of consciousness. Vishnu is the deity associated with Mercury, and he is the god of cosmic intelligence, the preserver and pervader of the universe. Shukra is the Vedic name for Venus; it means "shining", and it is the divinity of fertility and reproduction.

Mangal or Kuja are the Vedic names for fiery Mars. Its main deity is Subrahmanya or Skanda, who is the general of the gods, the son of Pleiades (Kartikeya), the spear-holder, and the spiritual warrior.

Guru is the Vedic name for Jupiter, and he is the teacher of the Gods, who is known for his divine intellect and speech. Shani is the Vedic name for Saturn. This male deity is the dark side of Shiva, the destroyer. The ascending and descending nodes of the Moon are called Rahu and Ketu.

In the Western tradition, Mercury gets its name from the winged messenger of the gods. He was also the god of thievery, commerce, and travel. Venus was the Roman goddess of love and beauty, so it is a fitting name for this brightly shining planet. The only objects in our Solar System brighter than Venus are the Sun and the Moon.

Mars is named after the Roman god of war. The planet got its name from the fact that it is the color of blood. Other civilizations also named the planets for its red color.

Zeus was the king of the gods of Olympus. The same god, Jupiter, was the Roman king of the gods. Considering that Jupiter is the largest planet in our Solar System, it makes sense that the planet was named after the most important god.

Saturn was named after the Roman god of agriculture and harvest. While the planet may have gotten its name from its golden color, like a field of wheat, it also had to do with its position in the sky. According to mythology, the god

Saturn stole the position of king of the gods from his father Uranus. The throne was then stolen by Jupiter.

The quest

Primitive rocket engines for propulsion were developed first in the 1920s. After the Second World War, in the 1950s and 1960s, there were rapid advances in the design of rocket engines and reliable launch vehicles, making it possible for satellites to reach orbit. Since then, humans have traveled into space, walked on the surface of the Moon, and have operated orbiting space stations where people can live for months. Humans in space require protection, as it is an incredibly harsh environment; it is a vacuum, with extreme temperatures and harmful radiation. Humans will likely return to the Moon in this decade, colonize Mars within another decade, and reach other planets in the solar system within another two decades.

To go beyond Earth, one needs to overcome its gravity that requires a velocity of at least 11.2 kilometers per second. By international agreement, space starts at 62 miles (100 km), which is called the Kármán line. The solar system's edge, called the heliopause, is about 18 billion kilometers from the Sun. The nearest star system to the solar system, Alpha Centauri, is approximately 4.37 light-years away, roughly 41 trillion kilometers, an incredible distance compared to distances in the solar system that are familiar from space exploration.

The solar wind, emanating from the Sun, creates a bubble that extends far past the orbits of the planets. This bubble is called the heliosphere, shaped like a long windsock as it moves together with the Sun through nearby interstellar space.

The heliosheath is the outer region of the heliosphere, just beyond the termination shock, the point where the solar wind slows abruptly, becoming denser and hotter. The solar wind piles up as it presses outward against interstellar space, and the interstellar wind, which is a stream of gas and dust flowing from interstellar medium, interacts with the solar wind.

The interstellar wind consists of partially ionized plasma. Because the Sun rotates once approximately every 25 days, the heliospheric magnetic field transported by the solar wind gets wrapped into a spiral. The solar wind affects the planets. For example, variations in the Sun's own magnetic field are carried outward by the solar wind, producing geomagnetic storms in the Earth's magnetosphere.

The boundary between solar wind and interstellar wind is the heliopause, where the pressure of the two winds is in balance. This balance in pressure causes the solar wind to turn back and flow down the tail of the heliosphere. It is a dynamic, ever-changing environment.

Early rocketry

Rocket engines that operate in space must be designed to work in a vacuum without any oxygen, so the spacecraft they power must carry fuel and an oxidizer, which together is called a propellant. In 1903 in Russia, Konstantin Tsiolkovsky proposed the concept of multistage rockets and suggested the possibilities of space travel. He is also famous for his memorable remark in a 1911 article: "The earth is the cradle of humanity, but one cannot forever live in the cradle."

In the U.S., in 1914, Robert Goddard received a patent for a rocket using liquid fuel and another for two-stage or three-stage rockets using solid propellant. His work influenced others working in rocketry, including Hermann Oberth and a young Wernher Von Braun in Germany, who was later to be part of NASA's ballistic missile, rocket, and spaceflight programs.

Goddard began experimenting with liquid oxidizer, liquid fuel rockets in September 1921, and successfully studied the first liquid propellant engine two years later. He now tested different variations to his design and flew the first successful liquid-propellant rocket engine in 1926. By the mid-1930s, his rockets were flying at supersonic speeds and reaching heights of nearly two miles.

Robert Goddard with the launching frame of the first liquid-fueled rocket, 1926.

Hermann Oberth wrote a book *The Rocket to the Planetary Spaces* (1923) that generated much interest in space flight amongst the public. Later, during 1928-1929, he served as a scientific advisor in Berlin for the film "Frau im Mond" ("The Woman in the Moon") and was given the assignment to build and launch a rocket as a publicity event just before the film's premiere. He built and fired up a liquid-fueled rocket engine in 1929.

By the late 1930s, powerful rockets had been developed as long-range weapons, and humanity entered the missile age as several countries started their own ballistic missile programs. The German V-2 rocket was the world's first long-range guided ballistic missile. In war, it was used primarily against targets in England and other parts of Western Europe. While thousands of V-2 ballistic missiles were used during WWII, they had poor guidance systems, and mostly missed their targets, which is in contrast with the accuracy of today's guided continental and intercontinental rocket systems.

In 1936, the American engineer, James Hart Wyld, developed the concept of a regeneratively cooled liquid rocket motor, which he named M-15. This used a double-hulled rocket nozzle that allows the rocket fuel to circulate and serve as a coolant. The innovative cooling design became the basis of all modern liquid-propellant rocket motors.

After WW II, engineers in the U.S. used the V-2's rocket engine components to help develop modern rocket programs. Oberth and his assistant Wernher von Braun were brought to the U.S. along with other scientists as part of Operation Paperclip. By 1948, they had launched a two-stage rocket to an altitude of over 322 km. After developing ballistic missiles for the U.S. Army, von Braun's group became part of NASA. This work was instrumental in developing NASA's human space program, including the moon landings.

The X-15 research rocket-powered aircraft is a hypersonic rocket-powered aircraft operated by the United States Air Force and NASA as part of the X-plane series of experimental aircrafts. The X-15 set speed and altitude records in the 1960s, crossing the edge of outer space and returning with valuable data used in aircraft and spacecraft design. The X-15's highest speed, 4,520 miles per hour was achieved in 1967. This is the highest speed ever recorded by a crewed, powered aircraft, which to this day remains unbroken.

From Sputnik to the Apollos

The late 1950s and 1960s saw tremendous technical growth in the field of astronautical engineering and rocketry. The Soviet Union's Sputnik-1 in 1957 became the first artificial satellite in space. Shortly after, the Soviets launched Sputnik 2, carrying a slightly heavier payload. It had four external antennas and broadcast radio pulses around the Earth. The success of the Sputnik launches came as a shock to the U.S., triggering the so-called "Space Race" between the U.S. and the Soviet Union. In 1960, Korabl-Sputnik 2 (Sputnik 5), with the dogs Strelka and Belka, was the first satellite with passengers to survive a trip into space.

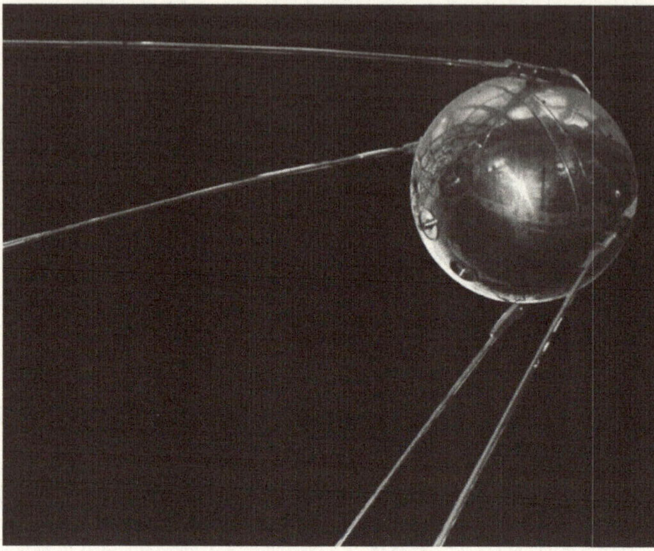

Sputnik-1, the size of a basketball, flew a low Earth orbit on 4 October 1957.

In 1958, the U.S. Government established the National Aeronautics and Space Administration (NASA) to ensure that US did not lag the Soviet Union in space exploration. This new organization soon launched its first satellite, Explorer 1.

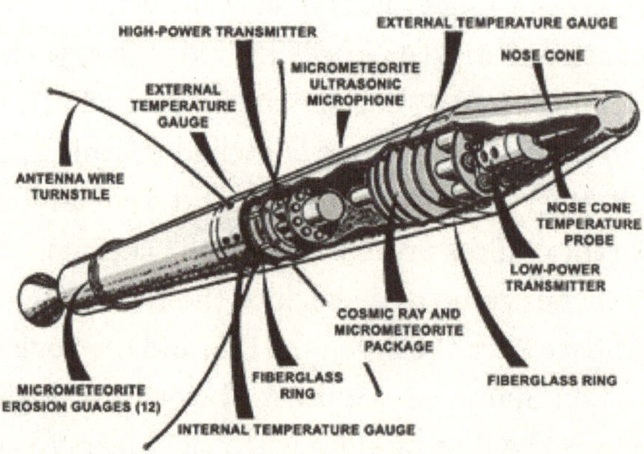

Explorer 1, the first satellite launched by the U.S. on 31 January 1958.

Following the Sputnik program, on September 12, 1959, Luna 2 became the first human-made object to impact the Moon, demonstrating early capabilities in interplanetary exploration. On April 12, 1961, Vostok 1 carried Yuri Gagarin, the first human to orbit Earth. The first woman in space was Valentina Tereshkova, who flew in June 1963 aboard the Soviet Vostok 6.

In May 1961, less than a month after Gagarin, the U.S. launched its first astronaut, Alan Shepard, on a suborbital flight. This flight was on the Mercury-Redstone (MR-3) rocket that Werner von Braun and the NASA engineers had developed. Shepard's Freedom 7 capsule reached an apogee altitude of 187km and a speed of 8,207 km/h before re-entering the atmosphere and parachuting into the Atlantic Ocean 484 km off the Cape. Alan Shepard was later to walk on the surface of the Moon as the commander of Apollo 14.

In February 1962, John Glenn became the fifth astronaut to fly in space and the first American to orbit the Earth completely, circling it three times during a five-hour flight. In 1963, NASA launched a series of communications satellites into orbit to demonstrate the feasibility of geosynchronous satellite communications from the Earth's surface.

The Gemini spacecraft program carried two astronauts in a capsule. Ten Gemini missions were flown between 1965 and 1966. The second piloted Gemini mission, Gemini 4, was in space for four days, and one of the astronauts performed a spacewalk.

Cutaway drawing of the Gemini space capsule.

In 1961, U.S. President John F. Kennedy laid down a national challenge to land humans on the moon before the end of the decade. There was a tragedy when Apollo 1 in 1967 had a fire in the cabin that killed the crew during a pre-launch test. The Apollo 8 mission proved the feasibility of leaving Earth and orbiting the Moon, followed by further lunar landing preparations with the Apollo 9 and Apollo 10 missions, however without landings.

The Apollo 11 flight on 16 July 1969 led to Neil Armstrong and Buzz Aldrin, Jr. successfully landing and walking on the surface of the Moon. After launching the three-stage Saturn V rocket, the spacecraft circled the Earth. The third stage allowed a translunar injection to send the spacecraft to the Moon. The next step was to undock with the third stage, dock with the Eagle lunar module, and pull it out, after which the stage was discarded to fly off into a solar orbit.

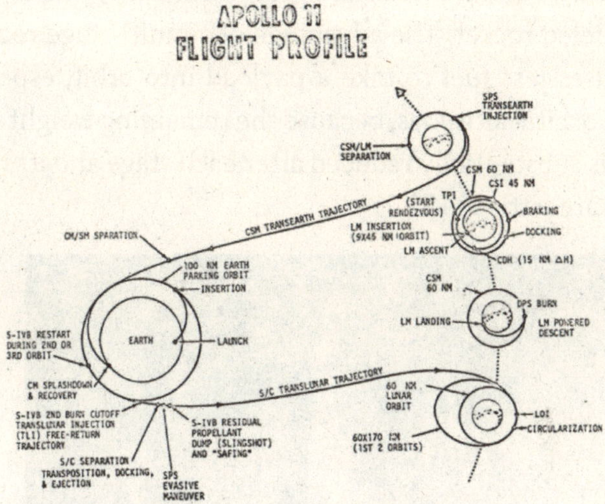

The lunar flight profile for the Apollo missions. Distances are not to scale.

After reaching the Moon, the spacecraft went into orbit, and the lunar module then undocked to begin the descent to the Moon, leaving Michael Collins to continue orbiting in the command module. Upon landing Neil Armstrong uttered the historic phrase "one small step for a man, one giant leap for mankind" as he took his first steps onto the Moon's surface on July 20, 1969. After less than a day on the surface, the two astronauts returned to orbit on the ascent stage with about 23 kg of rocks and other lunar samples. After a trans-Earth burn, the spacecraft made it back to the Earth in 44 hours, followed by a re-entry through the atmosphere, and eventually the command module parachuted into the sea.

Five further Apollo missions through 1972 (Apollo 12 through Apollo 17) sent astronauts to explore the Moon. The

Apollo missions were launched using a Saturn V, a three-stage liquid-fueled rocket. The advantage of a multi-stage rocket is that it uses less fuel to take a payload into orbit, especially for high orbital altitudes, because the remaining weight of the vehicle is substantially reduced after each stage and its rocket engines are jettisoned.

When fully fueled, the Saturn V rocket was 111 m tall and weighed 2.8 million kg.

The Soviet Union continued its advancements with Voskhod 2 on March 18, 1965, during which cosmonaut Alexei Leonov performed a spacewalk. The Soviet Union's space exploration efforts expanded further with Luna 9, which successfully achieved the first soft landing on the Moon on February 3, 1966, and transmitted pictures from the lunar surface. On August 17, 1970, Venera 7 successfully landed on Venus, providing data about the planet's atmosphere and surface conditions. Russia's continued involvement in space exploration was highlighted by the Soyuz MS-19 mission on August 9, 2021, which carried a film crew to the International Space Station (ISS).

Exploring the planets and the outer solar system

Starting in 1962, a series of Mariner spacecraft were sent from Earth to map the surface of Mars, Venus, and Mercury, and more missions continued to be launched until 1973. The Mariner program also achieved the first planetary flyby and gravity assist maneuver together with using solar sails to help propel the spacecraft into space. The solar sails represent a method of spacecraft propulsion using radiation pressure exerted by sunlight on large surfaces.

The planned Mariner 11 and Mariner 12 spacecraft were reallocated to the Voyager program and designated Voyager 1 and Voyager 2. In 1977, the two Voyager spacecraft were launched to send back detailed images of Jupiter and Saturn. Voyager 2 then continued to the outer planets Uranus and Neptune. Today, both Voyagers are still active, exploring the regions of space beyond the solar system, and sending back useful scientific data. Voyager 1 crossed the heliopause in August 2012 and continues to send data with Voyager 2 not far behind. In December 2024, NASA said that Voyager 1 was back operating normally after an unexpected blackout from the 47-year-old spacecraft that had traveled more than 15.4 billion miles from Earth.

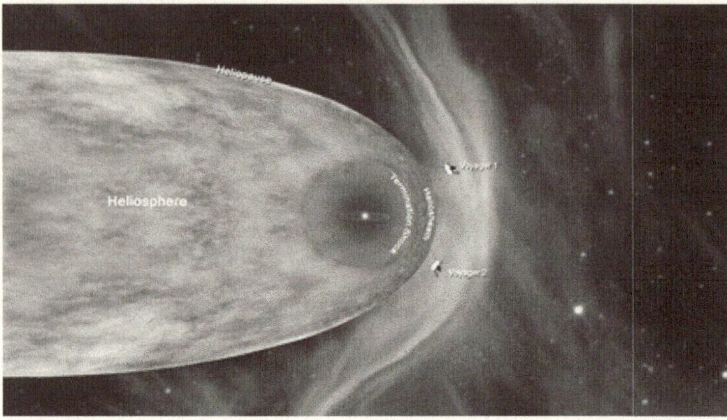

NASA's Voyager 1 and Voyager 2 outside the heliosphere, the region surrounding our star, beyond which interstellar space begins. (Credit: NASA/JPL-Caltech)

Japan's Aerospace Exploration Agency (JAXA) launched the world's first interplanetary solar sail spacecraft IKAROS (Interplanetary Kite-craft Accelerated by Radiation of the Sun) to Venus in 2010. It was the first true solar sail spacecraft fully propelled by sunlight.

Communication & earth observation satellites

While the 1970s saw a significant increase in the number of communications and navigation satellite launches, together with a substantial decline in manned space flights. By 1980, satellite communications became commonplace, and since then the capacity has expanded hugely to carry digital television and internet services to any home with a satellite receiving dish. Most communication satellites operate in geostationary or geosynchronous orbits, which are orbits with periods equal to the Earth's rotational speed (about 24 hours) at an altitude of approximately 35,780 km. Satellites provide real-time weather information, digital telecommunications, the internet, and GPS (Global Positioning System) navigation systems that allow one to pinpoint their location within a few feet of anywhere on the planet. GPS is now included in almost every automobile, smartphone, and smartwatch.

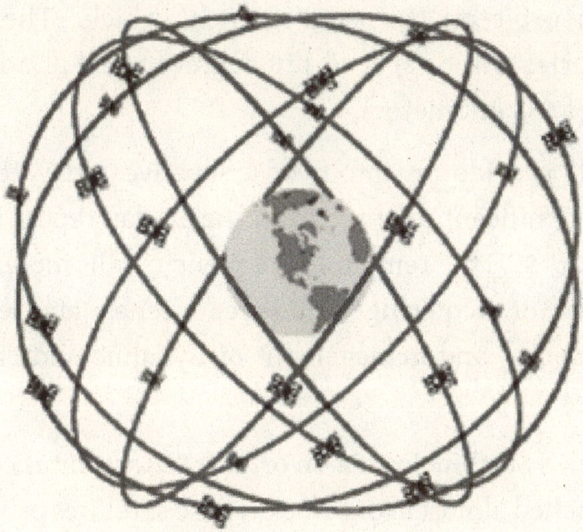

The specific orbits of the Navstar GPS satellites are designed to ensure that at least four satellites are visible above the horizon at any given time anywhere on the Earth's surface.

The 1970s saw the launch of Navstar satellites to support the Global Positioning System network. The first satellite to generate imagery records of Earth's surface, the Earth Resources Technology Satellite, was renamed Landsat 1 in 1975, with Landsat 9 launched in 2021. Landsat instruments have acquired millions of images that are publicly available through the U.S. Geological Survey (USGS). The available imagery has become a unique record of human activity on

Earth regarding urban expansion, deforestation, and melting of the polar ice caps.

Landsat Next, expected to launch in late 2030/early 2031, will represent a significant leap in measurement capabilities with improved temporal, spatial, and spectral resolutions. It will be a constellation of three identical observatories sent into orbit on the same launch vehicle. The triplet observatories will be spaced 120 degrees apart at an orbital altitude of 653 kilometers.

It will provide an improved collective 6-day temporal revisit, a significant upgrade from the 16-day repeat interval of Landsat 9. The temporal frequency will increase the probability of acquiring cloud-free scenes and enhance the monitoring and management of dynamic and changing landscapes.

SpaceX's Starlink low Earth orbit (LEO) satellites continue to be launched almost every week. These satellites orbit about 550 km above the Earth's surface compared to geostationary satellites, which orbit at about 35,786 km. The lower orbit reduces the time it takes for signals to travel between the satellite and the Earth. This provides advantages such as lower latency, greater coverage, and higher bandwidth, making them ideal for high-speed internet, particularly in remote and underserved areas.

The Space Shuttle

The Space Shuttle program, which began in 1981 and lasted until 2011, began a period of human space flight with reusable components and long-term residence in space. The central reusable part, the Orbiter, was launched with the help of two external solid rocket boosters that were jettisoned after about 90 seconds into the flight. The boosters then parachuted into the sea, were recovered, and reused. The Orbiter's flight to low-Earth orbit continued using its three main engines, followed by jettisoning the large external fuel tank. On returning to Earth, the Orbiter re-entered the atmosphere and landed like a supersonic glider on a runway.

Despite its engineering complexity, the Space Shuttle was a successful concept, providing valuable experience about long-term living in space. It was used for military and commercial purposes, including the deployment of communications satellites, interplanetary probes, various science experiments, servicing of the Hubble Space Telescope, and construction of the International Space Station and its subsequent servicing. Sadly, the Shuttle concept had two catastrophic failures, one during launch and the other during re-entry with the Orbiter.

The Space Shuttle launched various satellites for geographic, oceanic, and atmospheric studies, including weather forecasting. It launched the Hubble Space Telescope (HST) into low Earth orbit in 1990, and the telescope has remained operational for important deep space observations of stars in our galaxy and various astronomy research activities. Another space telescope, the Kepler Space Telescope, was launched in 2009 into an orbit around the Sun, and it provided much information about the universe, and led to the discovery of thousands of new planets.

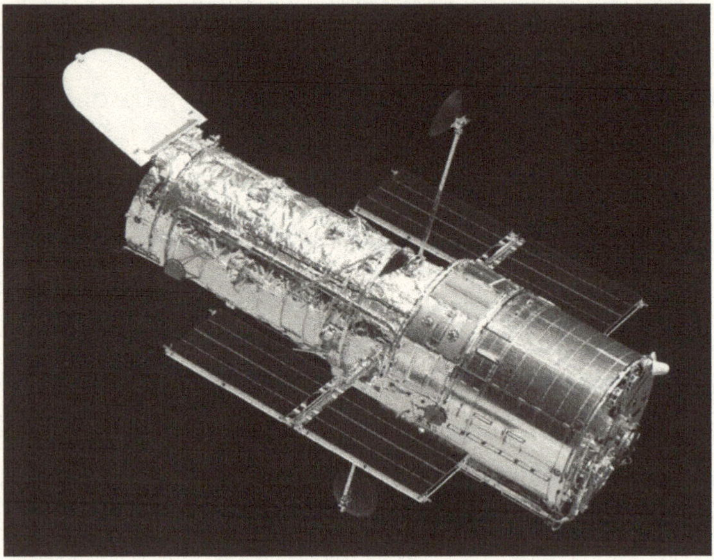

The Hubble Space Telescope (HST), which is in a low Earth orbit, remains a vital platform for astronomy research.

In December 2021, NASA launched the James Webb Space Telescope (JWST) into space. It can view objects that are too distant for the Hubble Space Telescope, providing information about the galaxy. The JWST is expected to make many future advances in understanding astronomy and cosmology, such as observations of the first star formations in the galaxy. The JWST operates approximately1,500,000 km beyond Earth's orbit around the Sun compared to the Hubble, which orbits only 550 km above Earth.

The James Webb Space Telescope (JWST), launched in 2021, operates 1.5 million km beyond Earth's orbit around the Sun.

In early 2024, JWST detected a galaxy called HD1 that existed about 330 million years after the Big Bang. This discovery pushes the boundaries of our understanding of the early universe, revealing how galaxies formed and evolved in their infancy. Additionally, JWST's advanced capabilities allow it to peer into the atmospheres of exoplanets with unprecedented detail. It has detected signs of carbon dioxide and water vapor in the atmospheres of exoplanets, helping scientists understand the conditions that might support life.

Space stations

In 1971, the Soviet Union launched the first space station, Salyut 1, to test the feasibility of astronauts being able to rendezvous with a space station and conduct scientific research, but it failed. In 1972, Skylab was launched by a Saturn V rocket and became the second space station to orbit the Earth. Skylab was very successful, and for about six years, it allowed visiting astronauts to perform scientific experiments.

A photograph of the ISS over the Earth. The ISS is a microgravity and space research laboratory that a variety of experiments.

The International Space Station (ISS) is a habitable satellite parked in a low Earth orbit. The Space Shuttle launched the first ISS components in 1998, which consisted of modules, external trusses, solar arrays, and other elements. It is a microgravity and space environment research laboratory occupied by astronauts since November 2000. The ISS flies in Low Earth Orbit (LEO) with an orbital period of just over 90 minutes. The ISS is big and low enough to be seen from Earth with the naked eye on a dark night; the reflecting solar panels look like a bright star that quickly arcs across the horizon.

Recently, astronauts aboard the ISS have been testing a new space habitat design called the Nexus module. This module includes expandable living spaces and advanced life support systems designed to enhance comfort and efficiency. It is a dedicated research laboratory space for life science experiments, where researchers can study the effects of microgravity on various biological samples; it is not a separate module itself, but rather a designated area within the station.

Space Launch System (SLS) and the Artemis Program

The NASA Space Launch System (SLS) is a heavy-lift vehicle under development since 2011. It lifted off for its flight debut on November 16, 2022, and this test flight propelled the Orion capsule, without any astronauts on board, on a round-trip mission beyond the moon, covering over 1.3 million miles and splashing down successfully back on Earth 26 days later.

In the long term, the SLS will become the primary launch vehicle of NASA's deep space exploration plans. It is a huge launch vehicle, 98 meters tall, comparable in size and weight to the mighty Saturn V. It holds 700,000 gallons (nearly 3.2 million liters) of propellant comprising cryogenic liquid hydrogen (LH) and oxygen (LOX). It has four first-stage RS-25 rocket engines, the same ones used previously on the Space Shuttle.

NASA's Artemis program aims to return humans to the Moon and establish a sustainable presence there, with ambitious plans extending beyond to Mars. In ancient Greek religion and mythology, Artemis is the goddess of the hunt, the wilderness, wild animals, nature, vegetation, childbirth, care of children, and chastity. In later times, she was identified with Selene, the personification of the Moon. She is the daughter of Zeus and Leto, and twin sister of Apollo.

The Artemis program seeks to achieve several significant goals. The program also includes the construction of the

Lunar Gateway, a space station planned to orbit the Moon. This station will be a staging point for lunar surface missions and a platform for scientific research and international cooperation.

The Artemis mission is to establish a long-term human presence on the Moon to study life support, habitat construction, and resource utilization. (NASA image.)

The Artemis I mission successfully tested the Space Launch System (SLS) rocket and the Orion spacecraft in an uncrewed flight around the Moon. Artemis II will be a crewed mission that will orbit the Moon, paving the way for the subsequent Artemis III mission, with aim to land astronauts on the lunar surface. The target for this landing is the Moon's South Pole, where water ice is believed to be present. This ice could be crucial for future lunar exploration, providing resources for long-term human missions.

Commercial space ventures

During the last decade, commercial and privately funded companies SpaceX and Blue Origin have developed launchers with a reusable first stage, which establishes the technical feasibility of reusing many components, including rocket engines. In addition, advances in computing and miniaturization of electrical and other elements have brought small satellites (i.e., those weighing 100 kg) to the forefront of commercial and research activities. The Planet Company is notable for its fleet of 3U CubeSats or nanosatellites, which provide daily global coverage in the visible spectrum with a ground resolution of approximately 15 ft (5 m). SpaceX has also begun to launch nanosatellites as part of a low-cost global communications network called Starlink.

Currently, three commercial spacecraft companies, United Launch Alliance (ULA), SpaceX, and Blue Origin are performing launches. ULA is operated by Lockheed-Martin and Boeing, providing spacecraft launch services to the Department of Defense and NASA. ULA operates expendable launch systems based on the Atlas and Delta launch system families, which have been used for decades to carry a variety of payloads, including telecommunications satellites and probes for interplanetary exploration. Blue Origin, owned by Jeff Bezos of Amazon, has had recent success in launching the New Glenn spacecrafts.

SpaceX aims to create technologies to reduce transportation costs and enable the colonization of outer space. It has developed the reusable Falcon series of launch vehicles, the first stage of which can be landed back at the launch site. The Falcon 9 uses the remaining fuel to reignite its engines in a series of decelerating burns to return to Earth. These burns help adjust the rocket's speed and reorient the vehicle into the proper position to enter the Earth's atmosphere and toward the final landing spot.

The SpaceX Dragon spacecraft was initially designed to supply the International Space Station (ISS) with cargo. Its crewed version, Crew Dragon, has since expanded the scope of human spaceflight. The first operational mission, Crew-1, launched on November 16, 2020, transported four astronauts to the ISS. Another milestone was achieved on September 16, 2021, with the Inspiration4 mission, the first all-civilian mission to orbit. Over three days, the crew conducted scientific and medical experiments and engaged in public outreach activities, highlighting the accessibility of space exploration to non-professional astronauts.

SpaceX has recently started developing the "Starship HLS" (Human Landing System) variant for NASA's Artemis program. This version of Starship is designed to land astronauts on the lunar surface as part of the Artemis missions. Overall, this program is focused on establishing a long-term human presence on the Moon, which involves developing new technologies for life support, habitat construction, and resource utilization.

SpaceX's Polaris Dawn mission launched on September 10, 2024 reaching a highly elliptical orbit with an apogee of approximately 1,400 km, the farthest any human has traveled from Earth since NASA's Apollo program. Its spent nearly five days in orbit, obtaining valuable data on human exposure to space radiation in their passage through the Van Allen radiation belt.

SpaceX has also developed the Falcon Heavy, the world's most powerful operational launch vehicle. Its maiden flight in February 2018 demonstrated its extraordinary capabilities.

The Falcon Heavy can lift nearly 141,000 pounds (64 metric tons) into low Earth orbit, more than twice the payload capacity of the next closest operational vehicle, the ULA Delta IV Heavy. Its first stage consists of three Falcon 9 nine-engine cores, with 27 engines generating over 5 million pounds (2,300 metric tons) of thrust at liftoff. This immense power enables Falcon Heavy to support a wide range of missions, from launching heavy payloads to deep space exploration.

The New Glenn rocket of Blue Origin, which stands over 320 feet tall, is one of the largest space vehicles ever constructed. This positions Blue Origin as a potential competitor to SpaceX in the realm of orbital launches. Industry experts anticipate that Blue Origin's entry into the commercial space sector will foster healthy competition, drive innovation, and potentially reduce costs.

The SpaceX Starship stands 120 meters in height and has a liftoff weight of 5,000 tons; it is a two-stage-to-orbit launch vehicle. The first stage, known as Super Heavy, serves as a booster to propel the second stage, also named Starship, into space.

The first Starship launch, as well as the second, were unsuccessful, but the third launch was able to achieve a suborbital flight. Future Starship missions aim to achieve orbital flights, demonstrating the reusability of both stages, and will include objectives such as payload deployments, lunar landings for NASA's Artemis program, and eventually crewed missions to Mars. The iterative testing process remains at the core of SpaceX's strategy to revolutionize space exploration.

Catching a spaceship

In recent launches, a part of a space rocket has been successfully caught on its return to Earth. After the SpaceX Starship rocket took off from the US state of Texas, its bottom part, the Super Heavy Booster, separated from the rest of the rocket. The booster, which was travelling at a speed of more than 17,000 mph, now re-entered the atmosphere. It was slowed down and gently guided back down to the launch pad where it was caught by a huge pair of mechanical arms, whilst still in the air.

SpaceX Starship's Super Heavy Booster is being grabbed as it returns to the launch pad at Starbase near Boca Chica, Texas, on Jan. 16, 2025.

This achievement brings the company closer to its plan of being able to make rockets that can be re-used. The Ship part of the rocket, which is where equipment and crew will be located, fired up its own engines after separating from the booster.

Both the Super Heavy booster and Starship spacecraft are powered by Raptor rocket engines, which burn a combination of liquid methane and liquid oxygen (Methalox). A key design goal for both Super Heavy and Starship is complete reusability and ability to perform controlled landings. In a fully reusable configuration, Starship can have a payload capacity of 150 tons to low Earth orbit. When expended (not recovered), it could carry up to 250 tons to orbit. One of the unique design features of Starship is that it can be refueled in orbit. SpaceX plans to launch special tanker Starships to refill Starships already in low Earth orbit, so they have the propellant needed to reach the Moon and Mars.

Beyond the U.S. and Russia, the European Union, Japan, India and China, have plans for scientific space exploration. On October 15, 2003, Shenzhou 5 marked China's first crewed spaceflight, with astronaut Yang Liwei becoming the first Chinese citizen to orbit Earth. It furthered its lunar exploration with Chang'e 1, launched on October 24, 2007, which mapped the Moon's surface. On December 2, 2013, Chang'e 3 landed the rover Yutu (Jade Rabbit) on the Moon, achieving China's first successful soft landing on the lunar surface.

The European Space Agency (ESA) has also made notable contributions to space exploration. On March 2, 2004, Rosetta became the first spacecraft to land a probe, Philae, on a comet (67P/Churyumov-Gerasimenko), providing valuable data on cometary materials and behavior. ESA's Gaia mission, launched on March 9, 2014, aims to create a comprehensive 3D map of the Milky Way galaxy, offering detailed data on stars and their movements.

India's space achievements include Chandrayaan-1, launched on October 22, 2008. This mission made significant discoveries, including water molecules on the Moon. Launched on November 5, 2013, India's Mangalyaan (Mars Orbiter Mission) entered Mars' orbit on 24 September 2014, making India the first Asian country to do so and the fourth space agency to reach Mars. Chandrayaan-3 successfully landed on moon on 23 August 2023. It was the first country to touch down near the lunar south pole, at 69°S, the southernmost lunar landing.

Japan's contributions include Hayabusa, launched on September 24, 2003. This mission collected samples from the asteroid Itokawa and returned them to Earth, marking a significant achievement in exploration. The follow-up mission, Hayabusa2, launched on December 3, 2014, targeted the asteroid Ryugu and successfully collected samples to study the solar system's origins.

Air-launched rockets

Other types of space vehicles may be carried to relatively high altitudes using a conventional airplane and then released into flight, much like the X-15 was launched initially. At this point, the rocket engines ignite and power the vehicle upward and into space. An example of this less conventional launcher design type is the Pegasus.

A large white airplane with a smaller rocket-powered craft underneath it the Orbital Science (now Lockheed-Martin) Lockheed L-1011 Stargazer launches the Pegasus into flight, continuing as a rocket-powered vehicle until it reaches low Earth orbit.

The Pegasus rocket is an air-launched vehicle designed for delivering small payloads into low Earth orbit. Released from its carrier aircraft at an altitude of approximately 12,000 m, it uses three solid-propellant stages and an optional monopropellant fourth stage to boost into space. The first stage features wings that generate lift in the atmosphere, allowing for faster altitude gain before being jettisoned. Afterward, the vehicle operates as a pure rocket until reaching orbital speeds and altitudes. With a payload capacity of about 455 kilograms, Pegasus is suited for small satellite launches and specialized missions.

Virgin Galactic has developed an air-launched vehicle capable of delivering payloads of approximately 500 kg to low orbits. The system uses White Knight Two, a mothership that carries the spacecraft SpaceShip Two to a cruising altitude. At this point, the rocket separates, ignites its boosters, and ascends out of the atmosphere. Two SpaceShipTwo vehicles, Enterprise and Unity, have flown missions. After 13 flights, these spacecrafts were retired, marking the end of an era for Virgin Galactic's early efforts.

The company has since shifted its focus to next-generation Delta-class space vehicles, which are designed to improve performance, increase passenger capacity, and reduce turnaround times between flights. Test flights for the Delta-class spaceplanes are expected to begin in 2025, with commercial service anticipated by 2026.

Deep space & beyond

Ventures into deep space lie beyond the capability of current technology. At this point, it is impossible to send astronauts into deep space and get them safely back to Earth. The distances in deep space are enormous, way beyond anything in the solar system. Alpha Centauri, the Earth's nearest Sun-like star system, is located 4.37 light-years away from the Earth, or about 40 trillion km away.

NASA's New Horizons probe was launched in 2006. Most of the post-Jupiter voyage was spent in hibernation mode to preserve onboard systems, except for brief annual checkouts. On December 6, 2014, New Horizons was brought back online for the Pluto encounter, and instrument check-out began. On January 15, 2015, the spacecraft began its approach to Pluto. It did a flyby of Pluto in 2015, providing the first-ever close-up photos of the distant dwarf planet and its moons. After this, New Horizons continued its journey deeper into the Kuiper Belt, a region of icy bodies beyond Neptune. NASA has announced it is to extend operations for the probe until the spacecraft exits the Kuiper belt, which is expected to occur between 2028 and 2029.

Artist's concept of the New Horizons spacecraft during its planned encounter with Pluto and its moons. (NASA image.)

The OSIRIS-REx and Hayabusa2 missions represent significant milestones in space exploration, focusing on retrieving samples from near-Earth asteroids. NASA's OSIRIS-REx launched in 2016 to study asteroid Bennu, successfully collected samples in October 2020 and returned them to Earth in September 2023. JAXA's Hayabusa2, launched in 2014, targeted asteroid Ryugu, collecting samples during two events in 2019 and returning them to Earth in December 2020. Both missions set out to analyze the composition and physical properties of these carbon-rich asteroids, providing insights into the early solar system and the origins of organic compounds and water that may have seeded life on Earth.

The Parker Solar Probe, launched by NASA in 2018, is a mission to study the Sun's outer corona and investigate the solar wind and space weather. It is the closest spacecraft to the Sun, reaching within 6.1 million km of its surface and traveling at speeds of 700,000 km/h. Equipped with a heat shield, the probe endures extreme temperatures while collecting unprecedented data on solar phenomena. Key discoveries include insights into the solar wind's acceleration, magnetic field "switchbacks," and the Sun's coronal heating. As of January 2025, Parker continues to transmit valuable data, enhancing our understanding of the Sun's influence on the solar system and improving space weather forecasting.

Many challenges and considerations arise in venturing into deep space. These include the vast distances involved, the need for advanced propulsion systems, life support for long-duration missions, and the potential effects of prolonged space travel on the health of astronauts. It is after all, a fact that humans developed and evolved on Earth, after billions of years of solar system evolution. The time it would take to reach even the closest star at our current technological level is prohibitively long, unless one travels close to the speed of light that itself would be a major challenge.

Space tourism

Space tourism is rapidly evolving from a futuristic concept into a growing industry, driven by companies like SpaceX, Blue Origin, and Virgin Galactic. Virgin Galactic's suborbital flights have offered brief experiences of weightlessness and dramatic views of Earth. Blue Origin's New Shepard rocket aims to provide similar experiences with a reusable design. SpaceX's Starship rocket represents a leap forward with its capacity for orbital and potentially interplanetary travel, highlighting the future potential for more extended and ambitious space tourism ventures.

The significance of space tourism extends beyond the personal excitement of space travel. Economically, it promises significant revenue and job creation while driving technological advancements that could benefit other industries. Public interest in space exploration is expected to increase, inspiring future generations of scientists and engineers. Additionally, efforts to develop more sustainable technologies are underway to mitigate the environmental impact of space tourism.

Ensuring the safety of space tourists is crucial, and it requires rigorous testing and the development of advanced safety approaches. Building the necessary infrastructure, including spaceports and support services, is also essential for successfully expanding space tourism.

Human ingenuity and innovation have overcome the engineering challenges of reaching space. The ability to successfully engineer rockets and spacecraft has allowed humans to put satellites into orbit, go to the Moon, send probes off to study other planets, and explore the solar system beyond the Earth.

It is a reality that commercial space companies will continue to develop technologies to make spaceflight affordable and open to renewal. This is expected to drive space travel, enable scientific discoveries and give rise to the possibilities of living, working in space and establishing permanent habitats beyond the Earth.

Space Ship One was developed by Mojave Aerospace Ventures without government funding at a reported cost of US$25 million. It was designed to:

- Carry three humans (one of them a pilot) in a sea-level pressurized cabin.
- Be propelled by rocket from an altitude of 15 km to more than 100 km.
- Reenter atmosphere and shed kinetic energy in an aerodynamically stable configuration.
- Land horizontally on a standard runway.

During its test program, SpaceShipOne became the first privately funded aircraft to exceed Mach 2 and Mach 3, the first privately funded crewed spacecraft to exceed 100 km altitude, and first privately funded reusable crewed spacecraft.

On October 4, 2004, it won the US$10 million Ansari X Prize, by reaching 100 kilometers in altitude twice in a two-week period with the equivalent of three people on board and with no more than ten percent of the non-fuel weight of the spacecraft replaced between flights. It was retired after winning the X Prize.

Prizes like the X Prize and the lure of space tourism in the AI dominated world where people seek adventure beyond what Earth can provide will be major drivers for the continuing efforts to build new spaceships and interplanetary colonies.

3

Life on the Moon

Research into lunar samples from the Apollo missions included searches for signs of life or organic compounds, but no evidence of life was found. Scientific speculation suggests that the Moon might have had conditions conducive to life in its ancient past, particularly during periods shortly after its formation and during significant volcanic activity. These periods could have provided water and a temporary atmosphere, though this remains highly conjectural.

Recent studies and discussions, especially around the Artemis missions, focus on the Moon's south pole, where permanently shadowed regions might harbor water ice. This ice could be a resource for future lunar bases and might also preserve ancient organic material or even microbial life. The Artemis program aims to return humans to the Moon for sustainable lunar exploration, and for the development of habitats where astronauts could live for extended periods.

While there's little evidence of current or past life on the Moon, the discussion around lunar habitability opens broader questions about life in the solar system. The Moon's study could inform our understanding of life's potential on other celestial bodies like Mars or the moons of Jupiter and Saturn.[23] The Moon can serve as a steppingstone for deeper space exploration, where the search for life might yield more promising results, particularly on the icy moons of the outer solar system.

But setting up a permanent camp on the Moon is much more difficult than sending astronauts there for a few days and bringing them back to Earth. Unlike the Apollo astronauts, lunar settlers wouldn't be able to carry all their rations and resources on their rocket. Instead, they would have to make much of what they need to survive out of the extra-terrestrial resources. Although these resources are rather limited in our current understanding, with ingenuity they can be transformed into much more. The relative proximity of the Moon to Earth would provide opportunities to explore human presence outside the Earth with relatively low risk since individuals could easily return to the parent planet.

The first task is to make breathable air. Surprisingly, this is easy, as the lunar soil is 42% oxygen. Using heat and electricity, this oxygen can be harvested by robots. Already, NASA has developed and field-tested prototype robots on Earth capable of fulfilling this role.

Beyond the basic requirements of a sustainable supply of air, water and food, settlers would also need to consider a

---

[23] Spilker L. Cassini-Huygens' Exploration of the Saturn System: 13 Years of Discovery. Science. (2019)

long-term power supply and permanent shelter. Fortunately, the lunar soil comes to the rescue again for both. It contains almost all the materials needed to build solar panels – a limitless and sustainable source of energy.

With no clouds limiting efficiency, the only problem with using solar power on the Moon is surviving the 354-hour lunar night. Solar batteries can store this energy but are far too bulky to be brought from Earth on a rocket. This is why selecting a high point of one of the Moon's poles has been suggested as a potential base. There, the base could enjoy permanent sunlight (except rare and short lunar eclipses) and therefore have continuous power. It is also where the Moon's ice is located.

Though the Moon has no liquid water, in 2018 NASA confirmed it does exist on the surface in ice form. Rovers could find, drill and gather this ice. Settlers would use this water for drinking and extract the hydrogen and oxygen for rocket fuel. And they would also save some for growing food. Unable to bring tons of rich, fertile Earth soil with them, they must use lunar soil. This soil is dusty, but it can be improved by using nutrients that help retain water.

The Moon settlers would need to bring seeds and earthworms from Earth. These worms recycle organic matter and improve soil structure, making them pivotal in creating a sustainable lunar agricultural ecosystem.

When the settlers arrive, shelter would likely come in the form of inflatable or expandable structures they brought with them. Inflatables are lighter, less bulky and would provide more room than rigid structures. But they would also need to protect the occupants from space radiation, particularly galactic cosmic rays that can damage electronics and DNA as well as small, frequent meteorite strikes.

A thick layer of lunar soil would provide plenty of shielding from cosmic rays. Scientists have developed a technique to 3D print bricks from lunar soil, powered solely by the Sun. Their vision is to make an igloo out of these bricks and cover it with loose soil. The igloo would surround the inflatable habitat – not only protecting against space radiation, but also the constant bombardment of micro-meteorites that the lunar surface experiences.

A major challenge for human life on the Moon is its low gravity. A person stepping on a scale on the Moon, would read one-sixth the weight on Earth. Clearly, living a long time on the Moon may present major health issues.

The experiment of life on the Moon ultimately comes down to the feasibility of moving beyond Earth. If we can populate the Moon and thrive, this would establish that we can survive in extremely harsh conditions. Most likely, then, we can survive and prosper elsewhere, beyond the planetary confines of Earth, to Mars and to the satellites of the outer planets.

Without a successful colony on the Moon, if only for scientific experiments, venturing beyond the Earth will be difficult although in some ways habitation on Mars will be easier. The advantage of experimenting with forming colonies on the Moon is the close distance of the Moon to the Earth, allowing for materials to be transported.

We will learn a lot from trying to live on the Moon for an extended period. These experiences will be essential for humans to move out of the Earth environment to the outer confines of the solar system. Populating the Moon is not a luxury or curiosity of expansion, it is a necessity for humans to get ready for the future.

Eyes on the Moon

The Moon has much to tell us about the origin and evolution of the solar system from a study of its history, but also as an outpost for the Earthlings. A robotic telescope on the moon could peer deeper into the universe than the James Webb Telescope and perhaps even help find life on exoplanets.[24]

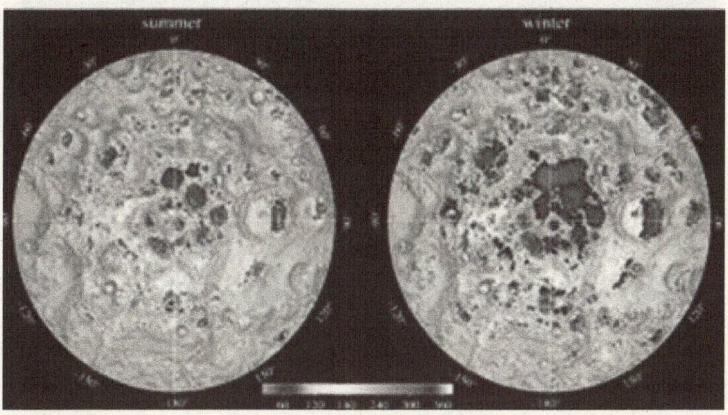

Permanently shadowed craters around the Moon's poles provide a perfect location for a next-generation infrared telescope. (Image credit: Jean-Pierre Maillard)

A robotic spacecraft mission in the Chinese Lunar Exploration Program of the CNSA, achieved a soft landing on the far side of the Moon with the touchdown in 2019 of

[24] Carroll, M. Envisioning Exoplanets: Searching for Life in the Galaxy. Smithsonian Books (2020)

Chang'e 4 rover. These advancements in autonomous navigation have laid the foundation for a lunar radio observatory that is built entirely by robots. As mentioned in the previous chapter, India's Chandrayaan-3 spacecraft touched down near the lunar south pole in 2023. It became the fourth space agency to successfully land on the Moon, after the US, the former Soviet Union and China.

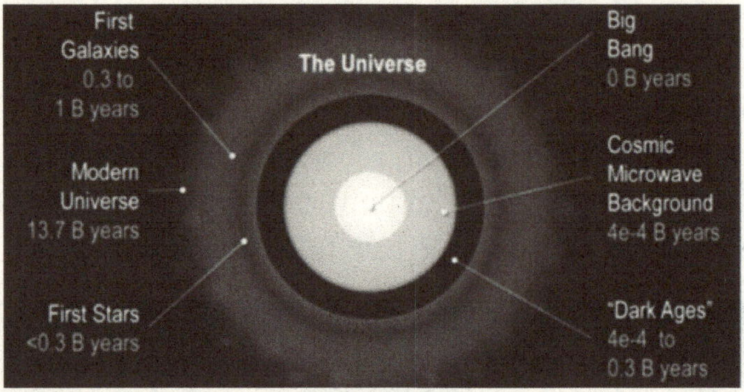

The history of the universe in a graph. Billions of years are indicated by the letter 'B'

A robotic telescope on Moon will enable the study of phenomena invisible to existing ground and space-based observatories. One idea involves placing a 42-foot-wide (13 meters) infrared telescope inside a permanently shadowed crater near one of its poles. It would mean a collecting area four times larger than Webb's and it would be more sensitive in the common 1 to 28 micrometers domain. Due to the cold temperatures on the Moon, a lunar infrared telescope could detect the far infrared part of the electromagnetic spectrum — radiation with wavelengths up to 200 micrometers. Longer wavelengths travel much better than shorter ones through the dust and gas that fills intergalactic space, bringing much information about the distant universe.

NASA by way of the Commercial Lunar Payload Services (CLPS) initiative is now funding lunar exploration. One payload is called the Radio Wave Observations on the Lunar Surface of the photo Electron Sheath (ROLSES). If successful, it would be the first radio telescope on the Moon and situated at the lunar south pole. ROLSES is to be placed there via the Intuitive Machines Nova-C lunar lander's IM-1 mission under the CLPS partnership.

Firefly Aerospace's Blue Ghost Mission-2 lander will carry the Lunar Surface Electromagnetics Experiment-Night (LuSEE-Night) to probe the *dark ages* of the early universe.This will be followed by the LuSEE-Night, slated to fly in 2026.

Designs for a lunar far side radio observatory include one with more than 100,000 dipole antennas and would become the most powerful telescope ever built for studying low-frequency radio waves. FarView could begin construction on the moon as early as 2030. Engineers working in tandem with robots would extract the vast majority of FarView mass from the lunar regolith, using these resources to make the most of the facility, including its power structure.

The far side of the Moon is shielded from natural and human-made radiation from Earth, which can interfere with observations of deep space. It could reveal new clues to some of the biggest questions in the history of the universe, including the origins of dark matter, dark energy, the formation of galaxies, star systems and planets that can sustain life. The observatory could also examine activity on the Sun's surface and measure the magnetic fields that may shield life on habitable exoplanets in our solar neighborhood.[25]

Beyond the relative safety of stellar systems, like our own, lie exotic conditions of the end stages of stellar evolution, nova explosions (which are likely double stars), supernova births and the formation of heavy elements, as well as stellar collapses to form neutron stars and the ultimate limit of the formation of black holes.

Back to the Moon, maria are ancient lava flows that erupted into the holes created by massive impact craters. One of the largest new maria identified on the farside was named Mare Moscoviense (MM), which translates to 'Sea of Moscow'. Tsiolkovskiy crater (T) is named after the Russian rocket scientist.

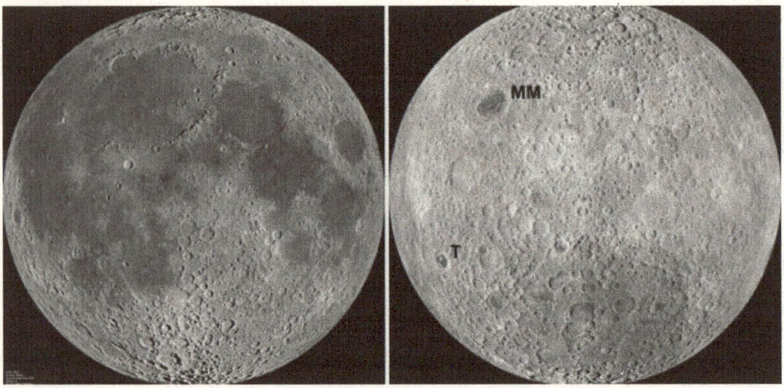

Moon images: the nearside (left, PIA14011) and farside (right, PIA14021) with the Lunar Reconnaissance Orbiter Camera constructed from many images taken in December 2010. MM stands for Mare Moscoviense and T is for Tsiolkovskiy crater.

[25] Summers, M. Trefil, J. Exoplanets: Diamond Worlds, Super Earths, Pulsar Planets, and the New Search for Life beyond Our Solar System. Smithsonian Books (2018)

Earth presents many difficulties for doing radio astronomy due to its thick atmosphere blocking a large portion of the radio spectrum. The proliferation of radio communication technologies has only made things harder. On the other hand, the Moon has no atmosphere and its nights last for weeks on end, which limits radio noise from the Sun. And the Moon also acts as a natural shield that blocks radio signals emanating from Earth.

Two main approaches to a lunar SETI observatory are being considered: an orbiter and a telescope on the surface. The basic idea behind a SETI lunar orbiter would be to scan for signals as it passed over the lunar farside and relay data back to Earth as it passed over the near side. One of the main advantages of an orbiter is cost, which is going down with the advent of commercial launch providers.

The proliferation of small satellites that are capable of accurate tracking combined with low-cost small launch providers like Rocket Lab means that a SETI orbiter could conceivably be sent to the Moon for less than $20 million.

But a SETI lunar orbiter would also come with a lot of compromises. It would only be able to conduct observations during the brief stretches when it was on the lunar farside, which would make a sustained observation campaign more challenging. The upshot is that an orbiter would have access to the full sky, whereas a telescope on the surface would be constrained by the Moon's rotation. The biggest downside of an orbiter is that it might lose a lot of the shielding benefits of the Moon and be more vulnerable to radio interference from Earth since it would be orbiting high above the lunar surface.[26]

One idea for a SETI observatory on the Moon is to use the naturally parabolic lunar crater as a radio dish, much like the Arecibo telescope in Puerto Rico and the FAST telescope in China, which are built into natural depressions in the land. The reduced gravity on the moon would allow for a radio telescope far larger than any on Earth, which will significantly enhance the sensitivity of SETI searches.

Another concept is for a radio telescope array that would deploy several smaller antennas across the lunar surface collectively forming a large radio telescope. One specific proposal seeks to deploy and operate an array of 128 dual-polarization, dipole antennas over a region of 100 square kilometers. The resulting interferometric radio telescope would provide unprecedented radio images of distant star systems, allowing for the investigation of faint radio signatures of coronal mass ejections and energetic particle events, and could also lead to the detection of magnetospheres around exoplanets within their parent star's habitable zone.

A farside crater observatory would depend on an orbiter to relay data back to Earth, creating radio interference for the dish, the very thing it was trying to avoid. It may not be necessary to start on the farside, however. It has been proposed to create a SETI observatory at the lunar south pole, the purported destination for NASA's first crewed return to

[26] McGarey, P. et al., How to Deploy a 10-km Interferometric Radio Telescope on the Moon with Just Four Tethered Robots. 2022 IEEE Aerospace Conference (AERO), Big Sky, MT, USA, 2022.

the Moon. While the south pole isn't as radio quiet as the farside, if an observatory was placed on the back of Malapert Mountain near the pole, it would also be protected from a lot of Earth's radio interference.

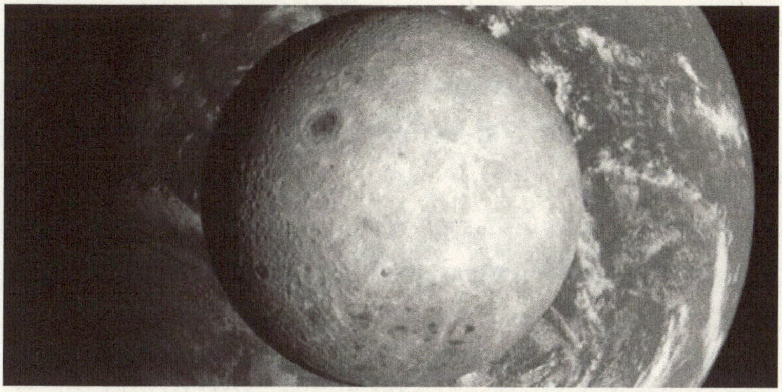

The farside of the Moon is an attractive place to carry out astronomy. NASA/Ernie Wright

Several types of astronomy would benefit. The most obvious is radio astronomy, which can be conducted from the side of the Moon that always faces away from Earth. The lunar farside is permanently shielded from the radio signals generated by humans on Earth. During the lunar night, it is also protected from the Sun. These characteristics make it probably the most "radio-quiet" location in the whole solar system as no other planet or moon has a side that permanently faces away from the Earth.

Radio waves with wavelengths longer than about 15 meters are blocked by Earth's ionosphere. But radio waves at these wavelengths reach the Moon's surface unimpeded. For astronomy, this is the last unexplored region of the electromagnetic spectrum, and it is best studied from the lunar farside.

Observations of the cosmos at these wavelengths come under the umbrella of "low frequency radio astronomy." These wavelengths are able to probe the structure of the early universe, especially the cosmic "dark ages" — an era before the first galaxies formed.

At that time, most of the matter in the universe was in the form of neutral hydrogen atoms. These emit and absorb radiation with a characteristic wavelength of 21 centimeters. Radio astronomers have been using this property to study hydrogen clouds in our own galaxy — the Milky Way — since the 1950s.

Because the universe is constantly expanding, the 21-cm signal generated by hydrogen in the early universe has shifted to much longer wavelengths. As a result, hydrogen from the cosmic "dark ages" will appear to us with wavelengths greater than 10 meters.

The farside of the Moon is a the perfect place to conduct low radio frequency observations of the early Universe's Dark Ages, as well as space weather and magnetic fields associated with habitable exoplanets.

Detection of radio waves from charged particles trapped by magnetic fields — magnetospheres — of planets orbiting other stars will be a significant application of farside radio astronomy.

This would help to assess how capable these exoplanets are of hosting life. Radio waves from exoplanet magnetospheres would probably have wavelengths greater than 100 meters, so they would require a radio-quiet environment in space.

Crater depths

The Moon also offers opportunities for other types of astronomy. Astronomers have lots of experience with optical and infrared telescopes operating in free space. However, the stability of the lunar surface may confer advantages for these types of instruments. Moreover, there are craters at the lunar poles that receive no sunlight. Telescopes that observe the universe at infrared wavelengths are very sensitive to heat and therefore must operate at low temperatures. JWST, for example, needs a huge sunshield to protect it from the Sun's rays. On the Moon, a natural crater rim could provide this shielding for free.

Permanently shadowed craters at the lunar poles could eventually host infrared telescopes. LROC/ASU/NASA

The Moon's low gravity may also enable the construction of much larger telescopes than is feasible for free-flying satellites. The cold, stable environment of permanently shadowed craters will have advantages for the next generation of instruments to detect gravitational waves — "ripples" in space-time caused by processes such as exploding stars and colliding black holes.

Astronomy will benefit from the infrastructure built up on the Moon as lunar exploration proceeds. This will include both transportation infrastructure — rockets, landers, and other vehicles — to access the surface, as well as humans and robots on-site to construct and maintain astronomical instruments. In the coming centuries, whole lunar cities could spring up, filled with factories and even hotels.

If FarView is built,[27] it would be able to detect some of the oldest light in existence. The universe began 13.8 billion years ago as a dense, fast expanding soup of matter and energy; around three hundred and eighty thousand years later, it had cooled enough for hydrogen atoms to hold together. After that came the Cosmic Dark Ages: millions of years without stars or galaxies although that understanding is changing.

By peering into the Cosmic Dark Ages, radio telescopes could help explain how the universe came to be. It is reasonable to assume that in the beginning matter was distributed evenly. To discover the logic behind the way

---

[27] Polidan, R. et al. FarView: An in-situ manufactured lunar far side radio array concept for 21-cm Dark Ages cosmology. Advances in Space Research 74, 528-546 (2024)

clumping of matter has taken place is a significant problem of cosmology.

Artist's impression of the LuSEE-Night lander on far side of the Moon.

A partnership of NASA and the Department of Energy, LuSEE-Night could possibly hitch a ride on ESA's Lunar Pathfinder mission.

A proposal known as FARSIDE (Far-side Array for Radio Science Investigations of the Dark ages and Exoplanets) would consist of 100 dipole antennas on kilometer-long tethers, rolled out by a small rover-like robot.

How big can a telescope or observatory on the Moon really be? In an idea named FarView, there will be a large, in-situ manufactured array of 100,000 dipoles, distributed over an area of 3 square kilometers, in the 84-kilometre-wide impact crater Pauli on the Moon's farside.

4

Mars and Planetary Moons

Mars has an axial tilt of 25.19°, that is quite like Earth's 23.44°. As a result, Mars has seasons much like Earth, though on average they last nearly twice as long because the Martian year is about 1.88 Earth years. The Martian soil is toxic due to relatively high concentrations of chlorine and associated compounds. Plants and animals cannot survive the ambient conditions on the surface of the planet.

The Martian day (or sol) is very close in duration to Earth's. A solar day on Mars is 24 hours, 39 minutes and 35.244 seconds. Conditions on the surface of Mars are closer to the conditions on Earth in terms of temperature and sunlight than on any other planet or the Moon, except for the cloud tops of Venus. However, the surface is not hospitable to humans or most known life forms due to the radiation, greatly reduced air pressure, and an atmosphere with only 0.16% oxygen.

Mars presents a hostile environment. The difference in gravity may negatively affect human health by weakening bones and muscles. There is also a risk of osteoporosis and cardiovascular problems.

The cardiovascular effects are complex. On Earth, blood within the body stays 70% below the heart, but in microgravity this is not the case due to nothing pulling the blood down, which can have several negative effects.

Dangerous amounts of radiation soak Mars despite it being much further from the Sun. Mars has lost its inner dynamo giving it a weaker global magnetosphere than Earth. Combined with a thin atmosphere, this permits a significant amount of ionizing radiation to reach the Martian surface.

There are two main types of radiation risks to traveling outside the protection of Earth's atmosphere and magnetosphere: galactic cosmic rays (GCR) and solar energetic particles (SEP). Earth's magnetosphere protects from charged particles from the Sun, and the atmosphere protects against uncharged and highly energetic GCRs. Building living quarters underground would significantly lower the colonists' exposure to radiation.

In the terraforming of Mars, novel engineering projects would be used to allow terrestrial life to survive free of protection or mediation. To be self-sustaining, a colony would have to be large enough to provide all the necessary living services.

The habitats will require energy production, industry, construction, and social and entertainment activities. They would also require equipment to produce food, propellant, water, energy and breathable oxygen.

A normal trip to Mars will require approximately nine months in space. Modified transfer trajectories that cut the travel time are possible and in use in robotic missions. With advanced spacecraft propulsion technologies, the trip time of forty days, or even two weeks, could be attainable.

The relatively strong gravity and the presence of aerodynamic effects make it difficult for heavy, crewed spacecraft with thrusters only, as was done with the Apollo Moon landings, yet the atmosphere is too thin for aerodynamic effects to be of much help in aerobraking and landing a large vehicle. Landing piloted missions on Mars would require braking and landing systems different from anything used to land crewed spacecraft on the Moon or robotic missions on Mars.

The one-way communication delay due to the speed of light ranges from about 3 minutes at closest approach to 22 minutes at the largest possible superior conjunction. Real-time communication, such as telephone conversations or Internet Relay Chat, between Earth and Mars would be highly impractical due to the long-time lags. But a satellite at the Earth–Sun Lagrangian point could serve as a relay during the blackout period to solve the problem. Alternatively, a constellation of communications satellites could be used.

Establishing power, communications, shelter, heating, and manufacturing basics will begin with robotic systems as a prelude to crewed operations. In the Mars transportation infrastructure envisioned by SpaceX, several robotic cargo missions would be undertaken first to transport the requisite equipment, habitats and supplies.

Solar energy is a candidate for power for a Martian colony. Solar insolation (the amount of solar radiation that reaches Mars) is about 42% of that on Earth, since Mars is about 52% farther from the Sun and insolation falls off as the square of distance. The thin atmosphere of Mars would allow almost all that energy to reach the surface as compared to Earth, where the atmosphere absorbs roughly a quarter of the solar radiation. Sunlight on the surface of Mars would be much like a moderately cloudy day on Earth.

It has been proposed to set up a first base at a Martian pole, which would allow access to water. Caves would provide a degree of insulation from Martian hazards for humans on the planet. These hazards include radiation, impactor events, and the wide range in temperatures on the surface.

Within our own solar system, Mars continues to be of interest due to its past environmental conditions that might have supported life.[28] While current conditions are harsh, the search for past or present microbial life forms continues, focusing on water ice, subsurface water, and ancient riverbeds. Past water on Mars means that conditions enabling life potentially existed in the distant past and as such they can be brought again together in the future.[29]

Scientists believe that the activity of volcanoes was a primary factor for the beginning of life on Earth. Volcanoes on Mars, including the giant Olympus Mons, and past active volcanoes that were active as recently as 150 million years

---

[28] Puumala, M.M., Sivula, O., Lehto, K. Moving to Mars: The Feasibility and Desirability of Mars Settlements. Space Policy 66 (2023)

[29] Harvey, R. The Solar System: Exploring the Sun, Planets and Their Moons. Amber Books (2022)

ago, indicate that life existed on ancient Mars. In contrast, the Moon never had active volcanoes so most likely it never had life.

Also, seemingly ancient Martian riverbeds, now dry, were active in the past. It appears that in the least microbial or even multicellular life thrived and became extinct as ancient water on Mars disappeared. Once the right technological conditions are sustained on Mars, life can be reestablished and the desire to return to Earth in a short time will not be urgent.

The gravity on Mars presents challenges and person weighing 100 kilos on Earth would only weigh 38 kilos. Long-term living conditions on Mars would present similar challenges as on the Moon, although better as the gravity on Mars is larger than on the Moon.

It is curious that in 1950 Wernher von Braun wrote a novel titled *Project Mars* about humans visiting Mars where they discover that the Martians had an advanced technological civilization, and they lived in subterranean pressurized habitats with controlled atmosphere. The Martian government was directed by ten men, the leader of whom was elected by universal suffrage for five years and entitled "Elon." Two houses of parliament enacted the laws to be administered by the Elon and his cabinet.

The Upper House was called the Council of the Elders and was limited to a membership of 60 persons, each being appointed for life by the Elon as vacancies occurred by death.[30] Today we've got Elon Musk of SpaceX, who wishes for humanity to go to Mars.

[30] Von Braun, W. Project Mars. Apogee Books (2006).

Martian life as imagined by von Braun

The vision of Mars sketched by von Braun is most interesting for its insights from the perspectives of both psychology and physics, and it will be sketched in the following pages in his own words without cluttering them with quotation marks.

On Mars, the Elon chose historians, churchmen, former cabinet members or successful economic leaders who could offer lifetimes of valuable experience. The Council of Elders, however, had limited authority. Proposed laws could be approved or disapproved as presented, but no riders, amendments or alterations could be voted. The Council of the Elders could stimulate, suggest and test legislation. Its main purpose was not action, but rather to ensure the continuity of the basic thought on legislation, as also the protection of that thought.

A lower house, the Assembly of Deputies, was devoted to action, for here demands were made and wordy battles were fought.

The main distinction between the Martian Assembly and analogous terrestrial legislatures was the electoral system. Whereas parliaments on Earth are composed of representatives regionally elected, the Martian Assembly of Deputies consisted of representatives of certain occupational groups.

This peculiar electoral system originated in the structure of the entire Martian community which, despite the close relationship between the thinking and feeling of the individual and that of his Earthling brother yet differed profoundly from the community of Earth. The ancient culture had long since emerged from the age when the welfare of the inhabitants of any region depended upon their being well and powerfully represented in the machinery of government.

Thousands of years of civilization on Mars had permitted refined technology completely to remove all regional concepts. Racial prejudice, national and local patriotism had not existed since time out of mind. Conditions of life on Mars, vastly different from those of Earth, made it possible for the planet to be governed without any form of regional representation. It was possible to make any conceivable journey within the confines of the planet in less than four hours in the high-velocity subways.

Subterranean existence prevented Martians from developing attachment to local scenery or dwellings. Furthermore, normalization of certain ideas and desires, mass production of all consumer goods, and synthetics replacing organic and inorganic natural materials and foodstuffs, made everyone conform to the artificial landscapes of their habitats. This also abolished the distinction between naturally rich and naturally poor regions, so that there no longer existed rich and poor regional groupings with conflicting interests.

The aging planet offered its inhabitants very few natural riches, and they had found that their individual lots could not be improved by attempting to bring their own regions into the decision-making, either politically or economically. Their economy was so fully integrated that no specific area could be left behind.

Instead of geographical differences of opinion, political debates went on between the representatives of various branches of science, technology and administration concerned with maintenance and improvement of living conditions. Traders and transportation people would differ with sociologists; physicians could not agree with ventilation engineers; private industry would argue with government; employees had grievances against employers; and so on.

The ultimate result of these differences was a congress of professionals from each of the branches upon which the highly evolved society of Mars was dependent, and each branch was represented in proportion to its importance.

The visiting Earthlings were lodged in the finest hotel in the capital city called Ahla. It was located on the upper levels of the circular, subterranean city. Although daylight never penetrated their apartments, the luminosity produced by the Martian interior paint provided all the goodness of sunlight.

The visitors from Earth were invited to the underground office building which fulfilled the functions of a city hall. Here was displayed a model of the amazing, bright, cheerful catacomb in which they were to live for more than a year. It was a huge, circular, conical mass of drifts, stops, halls and galleries, some twenty miles in diameter and half a mile deep in full scale. The central shaft held forty round, open levels, one above the other, each about one thousand feet in diameter. Around them were ranged the spacious quarters of the Martian government and the administration offices of the important companies.

The tubes of the metropolitan high-speed transit system proceeded radially from each of these levels to the sections of the city devoted to dwellings. From whatever level he happened to be occupying, the homeward-bound Martian had but to walk around the circle of the spacious shaft level where he was and enter the station which served the sector to which he wished to go. One after the other, small, six-seat cars continued to pass the platform at a walking pace. Selecting an unoccupied car, the Martian would step into it and dial a three-letter combination on a device not unlike a telephone. The residence of each Ahla-dweller had such a letter combination.

The little car would continue to move slowly to the end of the platform from which unoccupied cars were switched back into the column passing the platform. Any car with one or more passengers and a dialed combination would fly down the open tunnel ahead of it to the circular gallery corresponding to the first letter of the dialed combination. Here it slipped into an elevator which carried it up or down to the level prescribed by the second dialed letter. Then the car would slide right or left within its sector until it stopped at the dwelling coded by the third letter. The Martian would step out onto the tiny platform in front of his home door, while the car returned automatically to the central station.

Should the Martian or some member of his family wish to make a visit somewhere in Ahla after the departure of the car, he had to push a button beside the doorbell. Within three minutes an automatic mechanism at the central station would dispatch the same or similar car, after which dialing the appropriate three-letter combination would take them to any desired destination on any level and in any sector.

There were no streets, automobiles, pedestrians, traffic lights, collisions, parking problems or crowded trains. Ahla suffered from none of the nerve-wracking imperfections of an Earthly metropolis. When a Martian family desired a walk, the button near the door would summon a car in which they could ride comfortably to one of the many beautiful subterranean parks in which grew the most amazing, never-wilting trees and shrubs, and through whose maze of branches the Sun shone by day and the stars by night.

A most immaculate cleanliness was the hallmark of the huge, underground metropolis. There was no such thing as dust, because the air was cleaned, refreshed and even given the qualities of different times of year by a central air conditioning system. Individual preferences as to temperatures could be catered to if that selected by the operators of the municipal plant failed to please. Each dwelling had its own supplementary heating and cooling system, which could be adjusted to suit the occupant. Water, videophone, news, purchases - everything was done by wall plugs, spigots or cabinets in which the articles desired were taken out of one of the walls.

Cooking and serving was a lost art in Ahla, because the huge catering firms of the city were able to display their wares by television and to accept orders at the simple push of an electric button. Whenever such a button was pushed, the dwelling letter-combination appeared upon the order board of the company patronized, at a moment corresponding to the passage of a given dish across the television screen. Electronic cookers prepared it in a flash and, together with the trimmings, it was rushed into vacuum jars which were in turn loaded, complete with utensils, into metal containers and forwarded to the diner via the pneumatic tube delivery system which connected all portions of the city. Nor did dish washing plague the lives of Martians, for the remnants of the meal were simply replaced in the containers and pneumatically returned to the caterers.

A Martian wise man tells the visitors from Earth: "Tens of thousands of years ago, there were battles and wars on Mars until there arose a planetary government. Then came the long flowering of a beautiful culture. Standards of living rose higher and higher, despite the increasing depredations of global drought and erosion. Fine arts progressed to unthought-of perfection. Production of consumer goods on a vast scale well-nigh leveled the difference between rich and poor.

"The extirpation of contrasts and of the asperities of life decreased those tensions which theretofore had supported the ebullience of our inner drives. Now Mars is the home of a peace-seeking and easeful race, reposing upon the deeds of their ancestors. Everything is well organized; too well, indeed, to breed adventurous thoughts or actions.

"You Earthlings too will enter a period of wonderful cultural development, once the political unification of your planets is complete - and I envy you. But you will also have to pay the piper. It is impossible to satisfy the demands of the millions unless you adopt methods of production which produce identical goods likewise by the million, and true mass production means not only the standardizing of the goods produced, but also the standardization of requirements and tastes.

"You've just seen a very tiny example of what represents our greatest tragedy. It presses upon us all like a horrible nightmare, and those of us who have any feeling are fighting tooth and nail against an ever-increasing, gray uniformity which besets our lives. That uniformity is, I fear, too powerful ever to be overcome.

"Throughout thousands of years our people battled for personal freedom. Usurper after usurper who attempted to suppress freedom in the name of his own conceit was overthrown, and a free political system was established which covered the entire planet and has provided an unheard-of degree of stability for more than five thousand years.

"Our technicians liberated us from bodily need. That's right... But their methods, the horrible uniformity with which they invest everything, has literally scorched our souls."

The Martian chemists electrolyzed a certain proportion of the water as it arrived from the polar regions by passing powerful electric currents through it and thus disassociating it into hydrogen and oxygen. Oxygen was piped to the air conditioning and pressurizing plants, and there was used to regenerate the air of respiration, a continuous process. Hydrogen was used in the steelworks, where it was passed over beds of incandescent iron oxide, combining with the oxygen in them. This converted the ferrous oxides into metallic iron, to be worked into structural and other steels by ensuing processes of varying sorts. Simultaneously, the hydrogen burned to water vapor by taking up the oxygen from the ferrous oxide, and the water produced by condensing the vapor returned to the cycle.

In the air conditioning and regenerating plants, only half of their function was completed by the addition of oxygen to the air of respiration, for the carbon dioxide produced by breathing had to be eliminated to an equivalent extent. This was done in the same manner as in Holt's space vessels, where the exhaust air was pumped into pressure air washers and sprayed with water. Under the high pressure of these sprayers, the carbon dioxide was absorbed into the water droplets, thus separating itself from such air constituents as were insoluble in water. The carbonized water was expelled from the bottoms of the pressure washers and then, when depressurized, it rendered up its carbon dioxide in bubbles. The latter was then piped to food manufactories.

Martian food manufacture was one of the more extraordinary chemical miracles, although it was more biochemical than chemical, in the strict terrestrial sense of the word. Here the mystery of natural vegetable growth was duplicated by artificial means in retorts.

The biochemical establishments made practical use of that marvelous process which manifests itself in the growth of every plant and which Earthly scientists, in their inability fully to explain it, have called photosynthesis, an empty and non-significant title.

Photosynthesis is the word designating the ability of plants to produce fat, sugar and protein for their nourishment from carbon dioxide, water and a few chemicals which they ordinarily extract from the ground, and they do this by the mere presence of light.

The Martians in their food factories utilized the simplest form of vegetable growth known in nature, namely microscopic algae. The water supplied to the factories was first heat sterilized to destroy alien and undesired germs. Then it was "inoculated" with a relatively small amount of algae and was enriched by the necessary nutritive materials, in the main, carbon dioxide, magnesium, nitrogen and phosphorus. Water thus prepared was passed slowly under batteries of powerful artificial lights within glass tubes which allowed the ultraviolet rays to pass freely.

The algae grew both in size and numbers by the mysterious process of photosynthesis, forming, in accordance with the modifications of the process by minor supplementary chemical additives, primarily fats, sugars or proteins. When the light had taken effect and the water flowed out of the transparent piping, the augmented algae growth was centrifuged from the water and taken to extraction presses where the useful products were recovered and sent on to refineries to become food for the Martians.

Electricity was the mainstay of the hyper-refined Martian technological setup. Only by electricity was it possible to replace the complicated interplay between plant and animal life, the atmospheric processes and the farms and industrial production which on Earth permitted civilized life to be maintained and improved.

It was, therefore, quite natural that electricity had attained an almost mythical quality on Mars, for it not only helped transmute the waste products of the lungs into rich and tasty foodstuffs and permitted air of respiration and structural steel to be manufactured from desert dust and polar snow, it kept the infinitely involved gearing of the entire civilization in rotation.

Hundreds of horsepower were invisibly at work for every Martian, day and night; and to them alone Martians owed the high state of their material welfare, which were, in effect, the air they breathed, the food they ate, and the clothes they wore. Electricity bore the voice and image of distant friends into homes whenever people might wish. By that horsepower people were borne to any point of the red globe within the space of a few hours. It kept in motion the huge productive machine whereby they were able to earn in a short four hours daily their portions of whatever joys material accomplishment could offer.

Such conditions naturally brought in their train the establishment of energy as a standard of monetary value. Since nearly all production was almost fully automatic, the number of kilowatts used in manufacturing any sort of goods became the determining factor of their price and valuation. More energy per person meant more riches for everyone. Thus, on Mars, kilowatt hours took the place of Earth's gold standard of values.

Each time a new generating station was built, the amount of money in circulation increased, but no danger of inflation was incurred by this continual increase in the currency. Industries operated by the new source of power increased in productivity in the same measure as did the currency in circulation, thus maintaining a well-proportioned balance between the values of goods and money.

Telescopes on Mars

To put a radio telescope on Mars requires that it be taken to Mars and then deployed there. Naturally, this will be considerably more costly and difficult than just launching it into solar orbit.

A telescope on Mars would have less gravity to deal with on Earth but would still have to contend with it. This means it will have to be far more heavily built than a telescope designed to operate in microgravity. So not only is it more expensive per-kg to get it where it needs to go, one would need to send heavier equipment.

Also, it would have to deal with Martian dust, which is electrostatically charged and periodically carried aloft in large dust storms, and with the Martian ionosphere, which will block certain bands and cause interference just like Earth's.

Additionally, a telescope on the planet will be restricted to viewing a fraction of the sky determined by its location and the rotation of Mars. It would only be able to make uninterrupted observations near the poles, and then only if located close enough to the poles, which would raise issues with the winter temperature extremes and power issues with long high-latitude winter nights.

Since the parts of the sky the Mars telescope can observe are not necessarily the parts its Earth counterpart could observe, this creates a major problem for interferometry. This disadvantage is shared on the Moon. However, the Moon has no ionosphere, and therefore it can to some extent serve as a shield against interference from the solar wind, which is important in making certain very specific types of observations.

If small nuclear reactors were available, they would deliver enough power to keep a Mars base running, and these could also conceivably be deployed for a radio telescope.

The views of the cosmos from Mars will complement those from Earth and the Moon. If one could deploy telescopes beyond Mars, on the satellites of the planets such as Jupiter, Saturn and the outer planets, one would obtain a kaleidoscope of three dimensional views, which would give us stereoscopic visions beyond anything we can presently imagine.

Future generations of humans may obtain clues of which we are not presently aware, leading into stereoscopic visions of the rich systems of planets and their satellites and nearby stars.

The information would provide information on, where to build new communities, how to grow human societies and have them flourish, how to assist and preserve life and evolution of life, and the survival of species. An integrated system of systems with different layers of units will be a collective treasure with multiple uses.

The communication network on Mars could take advantage of the location of Phobos, the inner and larger of its two moons, synchronously orbiting Mars, where the same face stays facing the planet at about 6,028 km above the planet's surface. The use of a space elevator, which is a cable anchored to the surface and extending into space, has been suggested for transporting material to the surface. A rocket launch would be needed to get the rocket and cargo to the beginning of the space elevator 28 km above the surface.

Enceladus

Beyond Mars, there are numerous planetary moons that appear to be suitable for life beyond the Earth. Europa and Enceladus[31], the moons of Jupiter and Saturn, respectively, are potentially habitable due to subsurface oceans. Life could flourish in half-frozen slush areas on Europa and Enceladus, within the subsurface saltwater ocean of Ganymede, underneath the methane and ethane rivers of Titan, and maybe in brines in the deepest craters of the dwarf planets Ceres and Pluto. The icy shells of the ocean worlds may even contain pores filled with liquid water, and perhaps microbes.

The Galileo spacecraft discovered that Europa might be venting thin plumes of water 160 kilometers into space. It also found that Jupiter's magnetic fields induced a current, indicating salty liquid water was present within the sphere. Europa is the solar system's smoothest object, suggesting its surface is remade by interior processes more frequently than most other worlds besides Earth.[32]

Ganymede, the largest moon in our solar system might contain several layers of rock, water and exotic high-pressure ice. Interactions between rock and water are fundamental to microbial diversity. It is the only known moon with its own magnetic field, which causes auroras akin to those on Earth. The auroras sway when Jupiter's magnetic field fluctuates, which is partial evidence for a large saltwater ocean.

---

[31] Schenk, P.M. et al. Enceladus and the Icy Moons of Saturn. University of Arizona Press (2018)

[32] Howell S.M., Pappalardo R.T. NASA's Europa Clipper—A Mission to a Potentially Habitable Ocean World. Nat. Commun. 11, 1311 (2020)

Enceladus, the snow-covered moon of Saturn, is considered amongst the top candidates for hosting life in the solar system.[33] It is now known that the moon's giant plume of ice grains is rich in organic compounds.

A recent study reports strong confirmation of hydrogen cyanide, a poisonous gas that's also a molecule central to the origin of life, within the plume of water vapor shooting from the moon's surface. This discovery takes the evidence for possible habitability a step further.

Life requires building blocks such as amino acids, and hydrogen cyanide is one of the most important molecules needed to form these compounds. Because its molecules can be stacked together in many ways, the study authors refer to hydrogen cyanide as the "Swiss army knife" of amino acid precursors.

There is evidence that Enceladus's Ocean, which is hiding below the moon's icy outer shell and supplies the plume, holds a powerful source of chemical energy. Unidentified until now, the energy source is in the form of several organic compounds, some of which, on Earth, serve as fuel for organisms.

The combination of carbon dioxide, methane, and hydrogen in the plumes was suggestive of methanogenesis, a microbial metabolism named for the fact that microbes produce methane as a byproduct. Methanogenesis is widespread on Earth and may have been critical to the origin of life on our planet.

---

[33] Bhatia, G.K., Sahijpal, S. Thermal evolution of trans-Neptunian objects, icy satellites, and minor icy planets in the early solar system. Meteoritics & Planetary Science. 52 (12): 2470–2490 (2017)

There are many chemical pathways to potentially sustain life in Enceladus's subsurface ocean. That's because oxidation helps drive the release of chemical energy. Scientists are still a long way from proving that life could have emerged on Enceladus, but it is clear how chemical pathways for life could be tested in the lab and then established there.

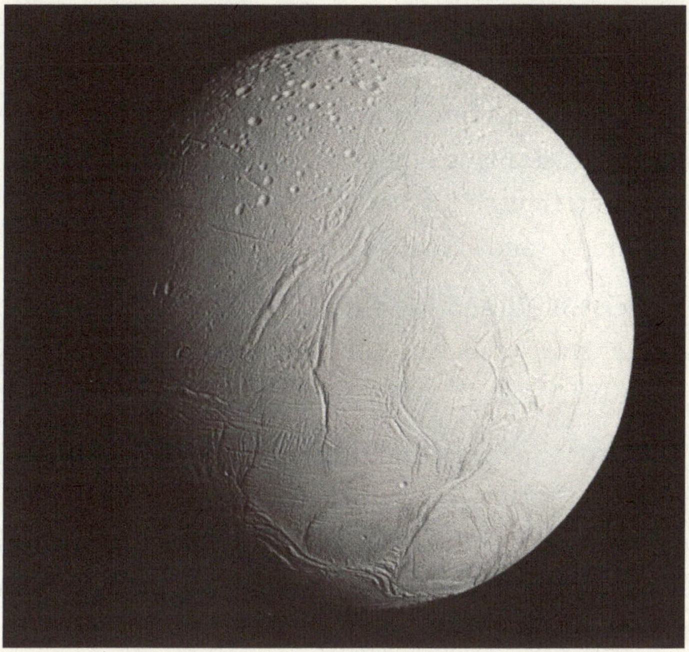

NASA's Cassini spacecraft captured this view as it neared icy Enceladus for its closest-ever dive past the moon's active south polar region.

Other planets

Outside our solar system, the universe contains vast numbers of planets. Planets of nearby stars can be studied for suitability of conditions for clues to life developing beyond the solar system. The planet, Gliese 12 b, discovered using NASA's TESS, orbits a red dwarf star and is particularly exciting due to its Earth-like size and temperature. It is located about 40 light-years away and orbits its star every 12.8 days. Its characteristics suggest it could be in the habitable zone, potentially allowing for liquid water, a key ingredient for life.

Another discovery from TESS is planet TOI-715 b, which is about one and a half times the size of Earth and orbits within what's considered the conservative habitable zone of its star. This means it might have the right conditions for liquid water, though further studies are needed to confirm its atmosphere and actual habitability.

Kepler-186f is another exoplanet in the habitable zone of another star. Its mass is 1.71 Earths, it takes 129.9 days to complete one orbit of its star, and it is 0.432 AU from its star.

The excitement around these planets stems from their potential to harbor conditions conducive to life, such as liquid water and suitable atmospheric composition. Many of these planets orbit red dwarf stars, which are cooler and smaller than our Sun. This makes finding Earth-sized planets easier due to the greater dimming effect during transits and the star's lower mass causing more noticeable wobbles.

These planets are not just potential new homes for life but also natural laboratories for understanding the conditions under which life can arise and persist over eons as it did on Earth.

Titan, Saturn's largest moon, presents a different set of opportunities and challenges. Its atmosphere is mostly nitrogen with about 3% methane. It is primarily ice and has oceans of ammonia rich water. It is the second largest satellite in the solar system, after Jupiter's Ganymede. Its thick atmosphere, which provides more pressure than Earth's, and the presence of liquid hydrocarbons (methane and ethane) in lakes and rivers, make it uniquely interesting. Titan's environment could theoretically support different forms of aviation or even boats, and its atmosphere might shield inhabitants from cosmic radiation better than Mars. The idea of colonizing Titan often revolves around its resources, particularly hydrocarbons, which could be used for energy or as raw materials for manufacturing. However, the extreme cold, distance from Earth, and the need for advanced life support systems due to its nitrogen-rich, oxygen-poor atmosphere pose significant challenges.

Mars offers Earth-like day-night cycle and the possibility of discovering past or present microbial life. Titan, on the other hand, intrigues with its Earth-like processes (like rain, rivers, and seas) but with different chemicals, offering a unique environment for scientific study and potential resource exploitation. Both Mars and Titan require overcoming significant technological hurdles. For Mars, this includes radiation protection, creating breathable atmospheres, and sustainable food production. For Titan, the challenges include dealing with extreme cold, utilizing its unique atmospheric chemistry, and managing the logistics of such a distant colony.

The idea of colonizing other worlds also touches on philosophical questions about humanity's future, our impact on other planets, and what it means to live beyond Earth. There's a narrative of pioneering and survival against odds, which resonates in both scientific communities and popular culture. It is worth examining whether Titan or perhaps Ganymede around Jupiter could evolve to Earth-like conditions for then it will be easy to populate them.

As Titan will be quite far from the expanding Sun, it might be a safe satellite to populate, even safer than the Galilean satellites of Jupiter and a future Titan warming environment would be friendly to life. Future generations will of course have to produce oxygen in its atmosphere. Titan would be safe from the expanding Red Giant, RG. As the RG sheds its outer layers into space, the Sun will turn into a compact and very dense White Dwarf, WD. The stability of the WD may allow a different collective experience, humans evolving to beyond what we now recognize as humans.

The current Sun is kept stable as thermal pressure balances gravity in the inner regions of the Sun. However, when the hydrogen fuel is used up in the center of the Sun and the outer regions of the Sun are ejected, the future WD will be stable because of quantum degeneracy pressure forced on electrons.

5

AI and the Future of Machines

There are many scientists and engineers who believe that computers will eventually become conscious. Given the fact that biological evolution is slower than technological evolution, there is fear that humans will be unable to compete with sentient machines. No wonder many prominent voices in the scientific and the tech worlds are claiming that artificial intelligence is leading humanity to a catastrophe.[34]

The question arises, what is going on? Can the "ghost of the self" in the body-machine be fully explained in terms of computation? Is it indeed inevitable that machines will become conscious unless Artificial Intelligence, AI, is strictly regulated? Some futurists and sci-fi writers are imagining a marriage of sorts between AI and brains so that, someday, technology will make it possible for humans to become "posthuman," transcending the limits of the human condition. There are others who believe that the only way to make sense of all the scientific facts is to take reality as a simulation[35], an idea that

[34] Kak, S., The Age of Artificial Intelligence. Garuda (2024)

[35] Greene, B. The Fabric of the Cosmos: Space, Time, and the Texture of Reality. Vintage (2005)

was used by Hollywood in movies, such as The Matrix and Watch Gamer. Another scenario is to imagine that once humans learn how to completely characterize brains, they will be able to copy themselves into computers, creating their emulations, or ems, in the process. Such ems could be replicated easily and so they will quickly outnumber real people.

In any event, if artificial intelligence evolves into conscious machines, that will surely mean the end of the world as we know it. Conscious machines will most likely take over the world and enslave humans, if not worse. Given this possibility, the prospect of machine consciousness is a most pressing challenge facing humanity currently.

Some may quibble over the idea of "consciousness" in machines. If we agree to use it in the sense of "awareness" and "personhood", machines with it will have access to greater autonomy than machines that lack it and, therefore, they would potentially replace humans in most jobs and raise insoluble problems of ethics and morality. Even if each human is provided with a universal basic income for food, shelter and entertainment, as is being proposed by many economists — and the machines leave us alone — it will be a dystopic life. This is surely beyond any science fiction world one could imagine.

There are many arguments in support of the idea of an eventual evolution of consciousness in computers. Since we cannot deny that the brain is a machine, then other machines with appropriate architecture should also achieve consciousness. Some object and say that brains, unlike classical computers, perform quantum operations deep in the neural circuitry and as such they are not really machines in the classical sense of machinery-based devices.

Now there are cognitive tasks during which the agent does not have a sense of subjective awareness. Such cognitive tasks are also being done by the machine, and it does so faster and with more reliability. One can imagine a future world where technology takes over and humans as well as other living forms play a secondary role.

Would this world cease to be a life-based world and change to one where technology reigns supreme? What would happen to moral, ethical and human-values? This AI-dominated world would be an alien world, right here on Earth, without the existence of alien extraterrestrial beings.

This is an interesting take on the evolution of life. It appears that what took billions of years for life to develop on Earth, might be upended by an evolution that is hard to predict even in the short term and totally out of our collective control.

Related to all these issues is the very idea of awareness-based, human-based existence, still very much anchored to biology. Awareness is something beyond computation for it is the ability to halt the processing in the brain machine at will to take stock of what is going on. Such a halting problem was considered by Alan Turing in 1936, and he showed that no algorithm can determine whether given a description of an arbitrary computer program and an input, the program will halt or continue to run forever.

If halting an arbitrary input at randomly chosen time is impossible from a computability point of view, and the aware mind does it, then one may conclude that *consciousness is not computable.*

In the naïve view, the world is an aggregate of physical objects and related forces, but this view has been shown to be wrong by quantum theory in which consciousness is complementary to materiality, and despite considerable efforts, it has not been possible to bring the observer within the framework of the theory. There are several arguments that show that consciousness is not computable, which means that it is not a property of matter.

Some theories assume that consciousness is an emergent phenomenon that arises when the complexity of the brain crosses a certain threshold. But if consciousness were due to complexity alone, then one can think of a system of sufficient complexity in which subparts of the system exceed this threshold with the result that a conscious agent has subparts that are conscious, which is impossible since consciousness is a unity. Awareness or consciousness cannot be physical, although our minds and their contents are definitely so. The split-brain phenomenon in which conscious unity is maintained in spite of dissimilar structures supports the view of a single ground of consciousness.

Much confusion in the academic approach to consciousness arises from its conflation with mind. The correct view is to assume that the mind is the apparatus used on the neural ground by which observations are made and experienced. If consciousness is a non-material category, then to speak of its evolution makes no sense. On the other hand, there has been an evolution of mind, and we can see a spectrum of minds across animal species, with each having its own distinctive mind. The computational capacity of the brain has a broad range and it is known that the human brain can have 25 independent cognitive centers, and this count is going to be different for other species.

In a document entitled the "Cambridge Declaration on Consciousness in Non-Human Animals," a group of prominent scientists asserted that the neurobiological structures needed to support consciousness are not uniquely human. They declared that according to convergent evidence, non-human animals have the capacity to exhibit intentional behaviors and humans are not unique in possessing the neurological substrates that generate consciousness. Non-human animals, including all mammals and birds, and many other creatures, including octopuses, also possess these neurological substrates. Recent work indicates that dolphins, elephants, magpies and domestic pigeons score high on cognitive ability.

Even fish possess neurobiological correlates and behavioral flexibility of sufficient complexity to support the hypothesis that they are capable of phenomenal consciousness. The insect brain also supports a capacity for subjective experience.[36] This is a significant issue to contemplate as insects outnumber all mammals by staggering numbers.

There are many examples of imaginative planning by birds and mammals, and more limited examples of similar ability of some fish, bees, and cuttlefish. Investigation of animal imagination can only be inferred for it lies in the private domain. Abstractions of language provide humans the unique ability to communicate the inner contents of their mind.

---

[36] Barron, A.B., Klein, C. What insects can tell us about the origins of consciousness. Proc. Natl. Acad. Sci. U. S. A., 113, eutio4900-4908 (2016)

Main propositions

According to the commonsensical view of reality, the world consists of nonliving, insentient matter and observers, who are living beings. Since our understanding of reality takes place in our consciousness, the knowledge aspect of reality is the more fundamental. To the extent, sentient beings are conscious, they have free will and therefore their future states cannot be fully predicted. However, since sentient beings also have physical bodies, the outcome, even if free, will be within the constraints of the physical system and so their behavior may be mapped probabilistically.

An insentient object can be fully described, assuming that a complete theory for it is available, and given its description and the current state, its future states can be estimated. Standard physics is about describing insentient objects and predicting their future states after interaction with other objects.

A sentient object has internal states that are not known to the outsider and, therefore, the future output states are not fully predictable, and this may be taken to be a result of free will. We will use the term "observer" for sentient object irrespective of their size since such an object is "aware" of its environment and takes that into consideration in its future action.

Observers are ignored in classical physics because it is assumed that their influence exists only at a local level, at small scales and there are other fields in chemistry, biology, and social science to account for the observers. It is generally accepted that big objects are not sentient.

To account for the ability of living forms to observe, note that the *One Observer Theorem* implies that there exists only one consciousness that transcends materiality.[37] Consciousness may be imagined as an additional dimension to reality that exists everywhere. Of the two aspects to reality—materiality and consciousness—the latter interpenetrates materiality. The reason there can be only one observer is because consciousness is a unity, and a conscious system or organism cannot have sub-parts that are conscious.

The capacity to observe is a consequence of the complexity of the brain. Therefore, the nature of observation in different animals or within different agents of the same species would depend on the size and structure of the brain and how it is being used. Major evolutionary transitions are marked in terms of qualitative changes in the way information is stored, processed, and transmitted. Examples of such transitions include the transition from single cells to multicellular organisms, and the transition from nonlinguistic communication to communication through symbolic language.

Different observers use the same ground of consciousness. Each observer is simply a different mind, which is the instrument of observation, and the same observer can be in one of many different mind-states.

The actual impression depends on the nature of the brain structure and how much of it has been recruited for the observation. The structures that correspond to specific cognitions can be vastly different. For example, birds don't

[37] Kak, S., No-go theorems on machine consciousness. Journal of Artificial Intelligence and Consciousness 10, 237-247 (2023)

have a cortex, yet their cognitive skills are on par with primates.

A cloak of probability covers the observation process, and there is no way to tell whether the observer made a free decision or whether it was a random event. This is the reason why causal science seems sufficient to explain reality. The implication of the action of consciousness at the macro level is hidden under the cover of probability.

Evolution of mind

It is interesting from the point of view of history of ideas that Indian scientific texts speak of evolution in the cosmological scheme of the Sāṅkhya school. The universe goes through cycles of creation and destruction. In creation, the evolution is from consciousness to akasha to air to fire to water and, finally, to earth. Physical structures arise by a process of aggregation that leads to greater complexity and likewise biological structures attain greater complexity with time. In this view, all organisms are descended, with modifications, from ancestors that lived long ago.

Darwin's theory made natural selection as the process that drove evolution for organisms with hereditary variations that rendered them better adapted to their environment than others. Currently an integrative approach to evolutionary reasoning, which incorporates the effects of variations in DNA, development, behavior, and culture is embraced by a growing number of biologists.

Variation in the hereditary units can arise randomly or can be partially directed because heredity and development can be coupled. Thus, stressful conditions during development can induce changes in gene expression that are transmitted to the next generation. There are multiple targets and levels of selection within individuals, between individuals and between lineages, and organisms have fuzzy boundaries. Organisms are not passive subjects of natural selection, and they actively construct the environment in which they are selected and bequeath these ecological legacies to their offspring. Clearly, consciousness will have a say in evolution, for it is reflected in the culture and the manner one responds to stress, and the choices that are made.

Here is a consensus list of multidimensional characteristics of the mind that has emerged from several academic disciplines:

- Binding/unification: seeing the fruit as a composite whole (green, round, smooth) yet with discernable features
- Perceptual richness: the ability to consciously perceive aspects of the animal specific environment
- Intentionality and agency: the mapping (representations) of body, world, action, and their relations and spontaneous activity and goal-directed behavior
- A sense of present: The present has a duration mapped to a working memory that allows data to be integrated
- Flexible evaluative system and goals: evaluating perceptions and actions as rewarding or punishing according to context

- A sense of self: registration of self/other and a stable perspective

Great apes appear to possess a theory of mind, allowing them to understand that others can have different beliefs based on their own knowledge, which was previously thought to be a uniquely human cognitive ability.

Even the octopus, an animal as different from to human as one can imagine, is sentient. Several parts of its behavior demonstrate a controlling mind interested in gathering information. Most octopuses are solitary, and many are cannibalistic, so they must always be on guard.

Driven by a combination of fear and exploration, they must survive by their wits. Their perceptual world is different from that of humans; they don't perceive color, but instead the polarization of light. They have a sense of self, and the ability to distinguish themselves from other organisms.[38]

Mirror of self-recognition

Awareness of the self rather than simply awareness of the environment is a sign of higher cognitive ability. This may be evaluated by the mirror self-recognition (MSR) test. In one experiment, a mark is covertly placed on the forehead or ear of the animal while it is briefly anesthetized. The animal recovers from anesthesia and is given a mirror sometime later. On seeing the image in the mirror, the marked animal spends time investigating the area of the mark and examining the fingers after touching the mark.

[38] Birch, J., Schnell, A.K., Clayton, N.S. Dimensions of animal consciousness, Trends in Cognitive Sciences 24, 789-801 (2020)

Primates that have passed the MSR test include chimpanzees, orangutans, and bonobos as have elephants. Humans begin to develop a sense of self and pass the MSR test starting around 18 months of age, and by 24–36 months, almost all children show a positive MSR response.

The cognitive abilities of some birds are now thought to be comparable to the abilities of some primates. These abilities seem to correlate with the relative size of the analogous avian prefrontal cortex. Indeed, the crow-like Corvidae (crows, ravens, magpies, rooks, jackdaws, and jays) appear to have the most advanced behavioral repertoire, as well as the largest prefrontal cortex (pallium).

Corvids are known for their ability to use tools. They can craft tools from sticks, leaves, or even wire to extract food from hard-to-reach places. Crows drop stones into a water-filled tube to raise the water level and access a floating reward, showing an understanding of displacement.

One can of course not know how the subject sense of consciousness differs across species. Although H. sapiens may have more advanced cognition, it is difficult to imagine that a human has a higher level of consciousness than a highly alert animal in pursuit of prey. It is likely that advanced symbolic processing in human cognition eclipses the subjective characteristics of experience.

We are arguing for considering consciousness and mind to be different and whereas evolution of mind has occurred in different species, there has been no evolution of consciousness. The action of free will by the mind is concealed within the randomness associated with the actions of the physical agent associated with it. All evidence points to mind being a product of specific structures in the brainstem and diencephalon in conjunction with neocortical connectivity and recurrent processing. Finding these structures in mammals and birds we can assert that their minds have similarity to ours.

The above evidence reveals that one cannot leave out the possibility that there may be further evolution in the biological realm that goes beyond the capacity of what humans can do now. At the macro level also the agency of consciousness may manifest itself variously so long as it conforms to the probabilistic structure of theories of physics. This agency is concealed within the probabilities associated with state reduction of the quantum state of various systems.

The Human-Machine Interface

The future of the human-machine interface (HMI) will enhance interaction between humans and technology in ways that are more intuitive, efficient, and integrated.

Natural Language Processing (NLP) will evolve, making interactions feel more natural, with machines understanding context, tone, and even cultural nuances. This could lead to AI assistants that are not just reactive but proactive, managing tasks autonomously based on learned patterns of behavior.

The technologies of Augmented and Virtual Reality (AR/VR) will redefine how we interact with digital environments. AR could overlay digital information onto the physical world, enhancing real-time decision-making in fields like manufacturing, medicine, or even daily life. VR might become a standard interface for immersive experiences, from gaming to professional training environments.

Brain-Computer Interfaces (BCIs) could revolutionize HMI by allowing direct communication between brain and machine. This could bypass traditional input methods, leading to applications in medical fields for paralysis patients, in gaming for more immersive experiences, or in enhancing human capabilities in various professional settings.

Gesture and haptic interfaces will likely become more tactile and responsive. Gesture controls, already seen in devices like the Kinect, could become ubiquitous, allowing for more natural interaction. Haptic technology will provide feedback that mimics physical sensations, making virtual interactions feel more real. Future interfaces will adapt to individual users, learning from their habits, preferences, and even emotional states. This personalization could extend to dynamically changing the interface's look, feel, and functionality based on the user's current task or mood.

As interfaces become more integrated with daily life, questions of privacy, security, and the impact on employment will grow. The development of these technologies will need to consider ethical implications, ensuring they enhance human life without compromising autonomy or privacy.

Future industry will focus on a more human-centric approach where technology and humans collaborate closely. This could mean machines taking over repetitive tasks, allowing humans to focus on creativity, problem-solving, and innovation, thus redefining work environments.

The trajectory of HMI is towards invisibility and seamlessness, where the interface itself becomes less noticeable, and the interaction feels more like an extension of human capability rather than a separate entity. This evolution will require ongoing dialogue between technologists, ethicists, and society at large to ensure that these advancements serve humanity's best interests.

Another area will be to use processing very different from the modules in the brain as in other species like octopuses. That might lead to new kind of perceptions. Cognition in a multi-species biological, evolving world appears to lead us back to the One Observer. If it is claimed that technology came out of biological evolution proceeding over billions of years, then how would AI-based evolution tied to advanced societies, proceed in the future?

It is also possible that the structures of contemporary society will become obsolete in the same way that city states were replaced by the imperium in ancient Rome: although cities continued to exist, effective power resided in the center of the empire, Rome, and ultimately in the person of the emperor. There are lessons to learn from human history in the West going back to the rise of Athenian democracy, the rise of kingdoms, that finally transformed into the Roman supremacy.

One can also investigate the origins of the "enemy" within the society and within the individual. Various philosophies state that inside each one of us is both friend and enemy. It is this division related to the left and the right, both within the individual and society at large, that leads to conflict and the movement of history.

It is realistic to imagine a scenario where AI-intelligence takes over from humans. Then the aliens are not "outside", they are just within, and they have intent to destroy human societies from within. Nations would become irrelevant and obsolete, quite as the city states of the ancient Greece yielded to the power of Rome.

The movement of the history of humanity appears to be leading to a new world, where technology is taking over and human intelligence is replaced by new, alien "beings" driven by AI.

The emerging "alien world" is of our own making and now its control may lie beyond our powers. If this turns out to be the case, then biological evolution over billions of years would become irrelevant in many ways.

To meet the threats of AI, we will have to go beyond often simplistic and binary ways that AI is being viewed. In other words, we must assess what the term itself means, what is artificial and what is intelligence. Once we come to an agreement, then the true issues associated with AI can be understood. It will require some basic agreements before pronouncements are made, otherwise arguments as to the role of AI will lead to nowhere.

AI-accelerated evolution

AI-accelerated evolution is the use of artificial intelligence to speed up or enhance evolutionary processes, either in biological systems or in computational models inspired by evolution.

In biology, AI can turbocharge something called directed evolution, a method used to optimize proteins or organisms for specific tasks (like making enzymes more efficient). Traditionally, directed evolution mimics natural selection by introducing random mutations and selecting the best performers over many cycles. AI steps in by predicting which mutations might work best, cutting down the trial-and-error time. For example, machine learning models can analyze vast datasets of protein sequences and fitness outcomes to guide the process, making it faster and more efficient than blind experimentation. Imagine evolving a protein that glows brighter or breaks down plastic—AI could slash the time from years to months or even days.

On the computational side, AI-accelerated evolution often ties into evolutionary algorithms. These are optimization techniques inspired by natural selection, where a population of potential solutions "evolves" to solve a problem—like designing a better neural network or finding the perfect shape for an airplane wing. AI can accelerate this by generating smarter starting points, predicting promising mutations, or pruning dead-end paths. There's even research where AI itself evolves, like programs that use evolutionary principles to build better AI models without human tweaking—think of it as survival of the fittest for code.

Researchers have used AI to simulate millions of years of evolution in a fraction of the time. A team recently leveraged a model called ESM3 to generate a new green fluorescent protein, mimicking 500 million years of natural evolution in a computational blink. This kind of work could lead to breakthroughs in engineering proteins to tackle real-world problems, like cleaning up pollution.

Thus, AI can push boundaries faster than nature or humans alone, opening doors to innovations we can barely imagine. But it is not perfect for predictions can miss the mark, and in biology, lab results don't always match the simulations. There is also an ethical angle in that speeding up evolution applied to living organisms may land humanity in a place for which it is not prepared.

6

Collapsing World Population and Future of Evolution

A generation ago, magazines and journals were full of dire warnings about runaway global population, a future of starvation and famine for millions, and population-induced climate change that will worsen the problem of feeding humanity and lead to chaos and societal breakdown.

The population of the Earth reached 1 billion in 1804, 220 years ago. It now exceeds 8 billion. Yet, unlike expectations of some decades ago that we were going towards an uncontrollable population explosion, the opposite is happening.

In other words, the reality has turned out to be different. The population is decreasing in many parts of the world, and it is expected to begin falling everywhere in the coming decades. There are scenarios that the world population a hundred years on may only be one-tenth of what it is now. If this were to happen, the world will change in unimaginable ways: whole regions would be depopulated, current political systems will be replaced, and in a world with very few children, the idea of the family will disappear.[39]

There are several reasons why this is happening. Technology has made birth control easy and artificial intelligence raises the specter of machines replacing humans at virtually all jobs so that parents do not want to have children who will have dim prospects in life. With the breakdown of the extended family, taking care of the child for many working parents has become unaffordable.

The total fertility rate (TFR) — the number of live children the average woman bears in her lifetime — has been falling since the 1970s. It has dropped under the 2.1 threshold (the "replacement rate," to account for infant mortality and sex imbalances), below which the population will fall, in more and more countries. This decline in fertility is perhaps the most remarkable trend of our times.

The decrease of the TFR below the replacement threshold is now well established as a general trend. In the US, the TFR fell below 2.0 in 1973, and in the UK, in 1974. In South Korea TFR was above 2.0 until 1984; in China until 1991. The current fertility rate in Iran is 1.6 and in India it is 2.0.

It takes a generation after TFR falls below 2.1 for the population to start tapering, and with another generation, the population collapse is in full swing.

If this is the case, then overpopulation is not the real problem, instead the real problem is underpopulation. This is quite a change of where we were heading and a scenario not anticipated even one generation ago.

---

[39] Kak, S., The Age of Artificial Intelligence. Garuda (2024)

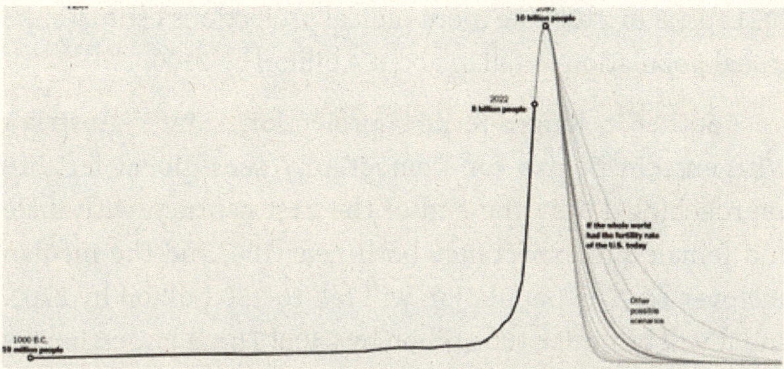

World population scenarios, New York Times

South Korea's current fertility rate is 0.68. This means that if the rate doesn't change across generations, a cohort of 100 Koreans will have 34 children, who in turn will have about 12, and in yet another generation it will be down to 4. In three generations (90 years), the survivors will be 48 in number of which 34 are 60-year-old, 12 who are 30-year-old, and 4 who are young. In another generation, their number will be further down to 16 of whom 12 are 60 years old and 4 are 30 years old.

Despite incentives to women to have more children, the South Korean fertility rate has kept on decreasing for the past 16 years. Demographers call it the "low-fertility trap" in which once a country's fertility rate drops below 1.5, it is virtually impossible to turn it around. Incentives have also been tried by France, Australia, and Russia with similarly disappointing results.

Although sub-Saharan Africa fertility rates remain well above the replacement rate, even in this region the fertility is expected to fall rapidly in the future. The global TFR, according to the UNPD's medium-variant projection, will fall from 2.3 in

2021 to 1.8 in 2100; the more radical projections estimate the global population to fall to about 4 billion by 2100.

Another longer-term projection by Austria's Wittgenstein Centre for Demography sees global fertility approaching 1.3 by the end of the 21st century, with male and female life expectancy both near 100, and the median age over 60. The population will fall to 250 million by 2200, and it will be under 100 million by 2300! These low estimates are not impossible given the transformative power of AI. This is a stunning change of events that in the most extreme case it could lead to a near-extinction of human society as we know it, after a mere 200 to 300 thousand years on Earth.

Projections of world population are based on assumptions on future mortality, fertility, migration, and other factors. Demographers leave out of their equation the fundamental changes in society due to the permanent disappearance of jobs caused by robots and AI machines and the impact it will have on the human psyche. In our view, the AI factor points to an even more drastic population decline than forecast by demographers.

The fragmentation of the traditional family, pervasive voluntary childlessness, the rise in single-parent homes, and the new normal of co-habitation and unmarried motherhood has made child-rearing very hard. We live in the age of narcissism where people are not thinking as much about raising children in extended kinship networks as about personal fulfilment and sense-gratification.

In East Asia, more women are choosing to marry later or not marry at all. Many Japanese youths show no interest in sex. There is a rise in living together outside of marriage, but illegitimacy and single parenthood are severely stigmatized. Only 2% of births occur outside of marriage, compared to 30–60% of such births in Europe and North America.

As populations shrink, the price of housing will fall. In Japan, the average value of real estate is less than half what it was in the 1980s and 1990s. There are more than 8.5 million *akiya*, or abandoned homes, in rural Japan; other estimates peg the number closer to 11 million. Spain has about 4 million empty homes, and the numbers in Italy are similar. This is the future for other countries as well.

One can also see national retirement systems becoming insolvent. Typically, pension systems take a portion of annual tax revenue and distribute it to retired people. This works fine when there are three or more working people for each retiree, for the state can tax one-fourth of the income and get three-fourths of one person's salary to distribute to the pensioner.

Japan is approaching one working age person for each pensioner and China will soon have one working age person for two pensioners, and clearly these ratios are unsustainable. The rest of the world will face the same problem soon.

The social reality on which modern political and economic arrangements rest is already facing severe pressures. As the population fall becomes extreme, current banking and political systems will be unsupportable.

One would have to consider that along with the entire human population of the Earth collapsing, different nations will face unique scenarios. The future will accelerate the disappearance of entire nations, not due to famine, war, or rapidly changing environment but because of the societal disruptions caused by plunging populations.

Clearly large population nations will "survive" as nations longer, while small population nations will become more rapidly extinct as small numbers will lead to collapse of everything associated with the very idea of a nation, including local and national governments, services, health and hospital units, local and national economies, cities and villages and even agriculture and food production. Small population nations will simply become untenable in every sense of the word and a thing of the past.

When we consider all factors, circumstances, projections, outputs of theoretical and experimental sciences, major issues emerge that sooner or later must be addressed, even if it takes many years of effort by individual countries, world organizations, corporations, health organizations and academia.

The question naturally arises; what world would that be? In a matter of a couple of centuries, if not sooner, the entire landscape of nations as stable entities will come to an end. The consequences of changes, whether welcome or unwelcome, will affect every level of existence on Earth.

There is no deeper, higher aspiration, more thought-provoking and meaningful aspiration than the vision of all beings on Earth forming an integral whole.

Replacement migration

Currently the population decline in rich countries is being countered by what the United Nations calls replacement migration. This migration is being encouraged because of the belief that the strength of a country is a function not only of wealth and military power, but also the size of the population.

Many elements go into bringing the migrants to the new shores. These include universities and legitimate businesses but also numerous fake colleges and criminal gangs who profit from human trafficking.

Replacement migration is changing the demographics of cities in the West, many of which are becoming white minority. For example, the proportion of whites in San Francisco, Chicago, New York, Los Angeles, and Houston is only 45%, 36%, 31%, 29%, and 24% of the population. In time, this process will extend to smaller cities as well.

Wholesale migration is an ancient phenomenon. Note that the original languages of Europe are not the currently dominant Indo-European family, and the remnants of the old languages are to be found in the Basque, Etruscan, and the Finnish language groups. There was a massive migration of the Indo-European people from the east into Europe in ancient times, perhaps in the third and second millennia BCE.

Likewise, what is modern day Turkey (Türkiye) is not the original homeland of the Turks, who began migrating into region only in the 11th century. This was originally the land of Greek, Anatolians, and other linguistic groups.

In China, a combination of war, famine, and disease during the Mongol conquests halved the population, compelling the Thai to move southward where they replaced the Khmer Empire. The Vietnamese, who defeated the Mongols, later subjugated the Chams (of the Kingdom of Champā) in the fifteenth century with the last remaining principality of Champā, Pāṇḍuraṅga (modern-day Phan Rang, Vietnam), capitulating in 1832.

The most dramatic replacement of population in history took place in the Americas and Oceania in the last five hundred years with the indigenous populations getting almost completely replaced by Europeans.

By 2021, the indigenous population in the US was down to about 3% of the total population, whereas in Canada, it is about 5%. The numbers for the indigenous people in Australia and New Zealand are about 3.8% and 16%, respectively.

Decline of European-origin population

Of Australia's 27 million population (2024 estimate) those of European ancestry are about 72%, with those of non-European extraction are 23%, and the Aboriginal Australian population is at 3%. The European figure for New Zealand's 5.2 million people is about 70%, together with 17 percent as Maori and 15 percent as Asian. In both these countries, whites will become a minority before the end of the century.

Canadian whites could drop to 10% of the population by 2100 and the ethnic white British could be as few as 33% by the end of the century.

The projected decline of European-origin people in these countries may be seen as a part of the large cycle of decline.

The process of migration is picking up, facilitated as it is by modern technologies of communication and transportation, and since the population of Europe is shrinking, almost all the migrants are from non-European regions. Today, more people than ever live in a country other than the one in which they were born. According to the UN, the global number of international migrants was approximately 281 million in 2020, with nearly 90 million of them living in Europe, with a similar number in Asia. In Bahrain, Kuwait, Oman, Qatar, Saudi Arabia and the United Arab Emirates more than half (53%) of resident populations are international migrants.

As people from the still high-fertility rate regions replace the native populations of Europe and the Americas, there will be conflict due to the disparity of their cultures and it may lead to civil war or partition.

For the United States, the Census Bureau provides race and ethnicity projections through 2060. The non-Hispanic white population is projected to decrease from 58.9% in 2022 to about 45% in 2060. The median age in the US is expected to increase to 47.9 years in 2100, up from 38.9 years in 2022. As the population ages, the Census expects there to be more seniors than children in the US in 2100.

This is part of a turning of history and there is nothing that can be done to reverse this process. As European people become a minority in the Americas and in Oceania, the nature of those societies will turn out to be very different from now.

Replacement migration is an important issue in the political dynamics in Europe and America. The left supports it because migrants tend to vote for parties that are for government control and benefits. The right promises to curb migration, but in the end does nothing as happened during the Conservative governments in Britain that let in more migrants than before.

Calhoun's Universe 25

A famous experiment of biologist John Calhoun over fifty years ago supports the idea that the size of a population doesn't depend only on outer prosperity. His colony of mice, once as large as 2,200 individuals, faded away and died when it lost the will to live.[40]

He took eight mice and put them in a large pen, with plenty of food and water and 256 separate apartments, which were accessible via mesh tubes bolted to the walls. The first pups were born 3.5 months later, and the population doubled about every 55 days (Phase B, explosive growth).

Eventually this growth slowed, and beginning at Day 315 and continuing for 245 more days, the population grew at a much slower rate, doubling every 145 days (Phase B, slow growth, stagnation). For some more time, the population continued to climb, peaking at 2,200 mice during the 19th month, and then it went on an inexorable period of decline and eventual death (Phase D, decline and death).

[40] Calhoun, J.B. Death squared: the explosive growth and the demise of a mouse population. Proc. Roy. Soc. Med. 66 (1973)

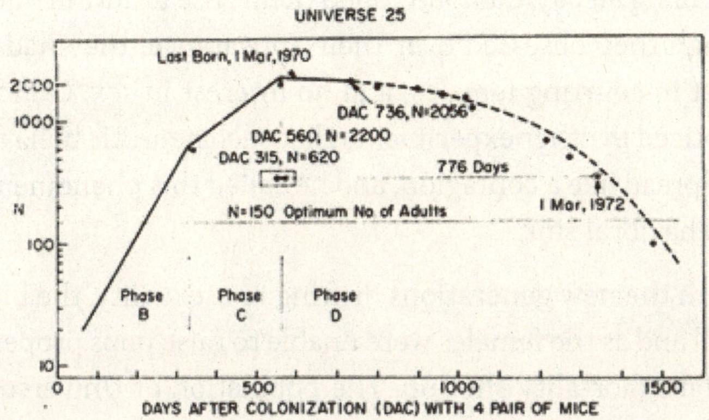

Life and death in Universe 25

Mice have social hierarchies, with dominant alpha males controlling harems of females. In the real world, the mouse who can't compete moves off to a distant place to start over. Since there was not a place to go, these losers that Calhoun called "dropouts" just gathered in the center of the pen and did nothing and indulged in senseless fights and violence. Even alpha males, who kept their harems in private apartments, suffered because of the frequency of challenges from hotshots and some got exhausted and stopped defending their places.

Apartments with nursing females were regularly invaded by rogue males forcing the mothers to fight back. The continuing violence was to the detriment of the young, and some mothers simply abandoned their places. Increasingly, mice in such conditions became neurotic, and many females began isolating themselves like hermits in empty apartments.

Many mothers pushed out their youngsters from the apartment early, before they were ready. Some mothers even panicked and attacked their own young. Neurotic males took to grooming all day. Calhoun called them "the beautiful ones." Although they obsessed over their appearance, they had no interest in courting females, and no interest in sex. Calhoun had noticed in other experiments that such neurotic behavior could spread like a contagion, and he called this phenomenon "the behavioral sink."

With the new generations drifting to asexuality, the birth rate fell and as the females were unable to raise pups properly, the infant mortality shot up. The population of Universe 25 began to decline. By the 21st month, newborn pups rarely survived more than a few days and soon new births stopped altogether.

By Day 1588, the inexorable decline had brought the population down to 27 (23 females and extremely old 4 males, the youngest of which was over 987 days old). Calhoun estimated that the last male would be dead on Day 1780.

The colony died even though it had plenty of food and comfortable accommodations. The fertility rate had fallen rapidly for psychological reasons and not for any adverse environmental factors.

Lessons for humans

Is the contemporary collapse of the fertility rate amongst humans like the one that took place in Calhoun's colony? If that is so, are we in the latter part of Calhoun's Phase C, with the danger of slipping into Phase D.

There are uncanny parallels of our story with human society. Both males and females are increasingly exhibiting neurotic behavior like Calhoun's mice. Examples of this are the widespread drug and social media addictions, homelessness, and gratuitous violence of American cities.

Western establishment's embrace of unrestricted illegal migration and DEI protocols diminishes white people, and it is like the abandonment of apartments and their young by nursing mothers in the mice colony.

Both males and females are falling into Calhoun's behavioral sink for social media amplifies mimetic desire as we do not autonomously decide on the objects of our desire; rather, we desire what others desire because we imitate their desires. This explains the contagions of transgenderism and narcissism.

The larger lesson that Calhoun drew from his experiment was that if a society does not offer opportunities for role fulfilment of sufficient complexity, violence and disruption of social organization will follow. "Just as biological generativity in the mouse involves this species' most complex behaviors, so does ideational generativity for man. Loss of these respective complex behaviors means death of the species."

The rejection of free thought, negation of biological sex in favor of fluctuating gender identity, gratuitous violence in cities, colleges, and schools in the United States indicates that we might be getting close to Phase D of Calhoun's Universe.

Death is hardwired into life. Modern societies are like Calhoun's Universe for they focus on physical welfare and don't incorporate an experience of death in life.

Life is not only morality and duties (D: *dharma*), material progress (A: *artha*), desires and enjoyment (K: *kama*), but also the experience of emancipation (M: *moksha*). It is this weaving together of freedom in our lives that inoculates us from *thanatos* (desire for death) and keeps our spirit alive. Western societies don't have liberation as a component of the template for life and individuals feel adrift and lacking in agency.

Artificial womb technology

The process of loss of interest in making babies as in Calhoun's colony could be mitigated by means of artificial womb technology (AWT) where babies are either partially or fully grown in an artificial womb.[41] At the present time, an artificial womb is an medical device intended to provide a womblike environment for extremely premature infants. These devices have been tested with hundreds of lambs (and some piglets), but it is not clear how this technology will work for humans.

A complete outside of the womb growth of an embryo to a baby is what has traditionally been called a "test-tube baby" and is formally called ectogenesis. Complete ectogenesis would surmount many forms of infertility, hence enabling biological parentage while avoiding the use of surrogacy. But it could also devalue pregnancy, and lead to commodification of pregnancy, babies and motherhood and a disconnect between sexuality and procreation.

In recent years early artificial embryos from mice stem cells have been created. The synthetic embryo does not require sperm, eggs, nor fertilization, and were grown from only embryonic stem cells or also from stem cells other than

---

[41] De Bie, F. R. et al. Ethics considerations regarding artificial womb technology for the fetonate. The American Journal of Bioethics, 23(5), 67–78 (2022)

ESCs. The structure had an intestinal tract, early brain, and a beating heart and a placenta with a yolk sac around the embryo.

No matter whether the AWT technology leads to ectogenesis, such research will have an effect on our consideration of the development of life and profound psychosocial implications for future societies.

The Future of evolution

Any interventions by humans that speed up biological evolution may be seen as a part of natural evolution for we do what we do impelled by forces greater than us.

The artificial environment we live in produces new selective pressures very unlike those we faced, say, in the ice age. We're poorly adapted to this modern world; it follows that we'll have to adapt, and that process has already started. As our diets changed to include grains and dairy, we evolved genes to help us digest starch and milk. When dense cities created conditions for disease to spread, mutations for disease resistance spread. And for some reason, our brains have gotten smaller.

To predict where this goes, we'll look at our prehistory, studying trends over the past 6 million years of evolution. Some trends will continue, especially those that emerged in the past 10,000 years, after agriculture and civilization were invented.

We're also facing new selective pressures, such as reduced mortality. Studying the past doesn't help here, but we can see how other species responded to similar pressures. Evolution in domestic animals may be especially relevant – arguably we're becoming a kind of domesticated ape, but curiously, one domesticated by ourselves.

The advent of CRISPR and other gene-editing technologies suggests a future where humans might deliberately choose traits for their offspring, leading to what could be called "designer babies." This could result in humans with enhanced physical capabilities, intelligence, or resistance to diseases.

Advances in medicine could lead to significantly extended human lifespans. This might not just mean living longer but also maintaining youthfulness and health for prolonged periods, potentially altering social structures, retirement, and generational dynamics.

The integration of technology with human biology might lead to cognitive enhancements. This could involve brain-computer interfaces or genetic modifications aimed at increasing intelligence, memory, or other cognitive functions. Such enhancements could redefine what it means to be human, potentially leading to a divide between those who can afford such enhancements and those who cannot.

As humans continue to alter the environment, future generations might evolve or be engineered to better adapt to new climates, pollution levels, or even space environments. This could include skin changes for UV protection, respiratory adaptations, or other physiological adjustments.

Evolution isn't just biological; cultural evolution could lead to significant changes in human behavior, social structures, and even physical traits through sexual selection. For instance, preferences for certain physical or behavioral traits might become more pronounced, influencing the gene pool over time.

As we already see, evolution will have to consider the advance of present-day AI and machine-based solutions to a host of health, social, agriculture and technological solutions, in the tapestry of a multitude of societal issues.

If humans colonize other planets or space habitats, evolutionary pressures in these new environments could lead to new human subspecies or adaptations. Gravity, atmosphere, and isolation could all play roles in shaping these future humans.

The line between humans and machines might blur further. Prosthetics, implants, and eventually, full integration with technology could redefine human capabilities, leading to a form of evolution where technology becomes part of our biological identity.

The ethical implications of guiding human evolution through technology are profound. There's a risk of creating a genetic underclass, where only the wealthy can afford enhancements, leading to societal divides. Conversely, there's an optimistic view where these technologies could be used to eliminate genetic diseases, enhance human potential universally, and lead to a more resilient species. Nevertheless, human progression, whether driven by natural selection or human intervention, will likely continue to be shaped by our interaction with technology, the environment, with all life forms and with each other.

Will human societies evolve for the general well-being of humanity, the well-being of natural ecosystems, the stability of the environment, the reduction of health risks, the stability of agrosystems, food production, stability of moving goods, products and humans across the globe, educational systems at all levels? Could it lead to the breakdown of everything imaginable? There is no reason this would be the case, yet our own very rapid evolution as social beings, our imprint on the environment, the threat of mass extinctions of human systems[42], the threat of local and global wars, all will have to be brought into the picture and examined in different contexts.

A major characteristic of human beings and human society is the ability to think collectively, to plan and to evolve, keeping the whole system in equilibrium. Central to everything associated with human beings are the functions of the human mind. The human mind is not a "thing" like a body part; it is full of abilities to address the most daunting issues that humans face.

Since the mind has evolved in the past, it is natural to assume that there will be further evolution in time and the capacity of the future mind may go beyond what humans can currently do. In fact, the very ability of the human mind to adapt, the ability to think and find solutions at all levels, the ability to think creatively, has potential for evolving beyond the present. Yet, this very mind has the capacity to devise destructive means that may threaten us at all levels.

---

[42] Brannen, P. The Ends of the World: Volcanic Apocalypses, Lethal Oceans, and Our Quest to Understand Earth's Past Mass Extinctions. Ecco (2018)

Lifespan

Humans will almost certainly evolve to live longer. Life cycles evolve in response to mortality rates, which depends on likely predators and other threats to life. When mortality rates are high, animals must reproduce young to ensure that their genes are passed on. There's also no advantage to evolving mutations that prevent ageing or cancer, for the animal will not live long enough to use them.

When mortality rates are low, the opposite is true. It's better to take one's time reaching sexual maturity. It's also useful to have adaptations that extend lifespan, and fertility, giving one more time to reproduce. That's why animals with few predators, such as animals that live on islands or in the deep ocean, or are simply big, evolve longer lifespans. Greenland sharks, Galapagos tortoises and bowhead whales mature late, and can live for centuries.

Even before civilization, people were unique among apes in having low mortality and long lives. Hunter-gatherers armed with spears and bows could defend against predators; food sharing prevented starvation. This delayed sexual maturity, and led to long lifespans that could go into the nineties. However, child mortality was high, and it approached 50% or more by age 15, and life expectancy is estimated to have been just 35 years.

Then, in the past two centuries, better nutrition, medicine and hygiene reduced youth mortality to under 1% in most developed nations. Life expectancy soared to nearly 80 in developed countries. Now, there's little need to reproduce early. If anything, the years of training needed to be a doctor, CEO, or carpenter incentivize putting it off. And since our life expectancy has doubled, adaptations to prolong lifespan and child-bearing years are now advantageous. Given that more and more people live to 100 or even 110 years - the record being 122 years - there's reason to think our genes could evolve until the average person routinely lives 100 years or more.

Good health is maintained by certain mechanisms that decline during aging although it is not known how it happens.[43] There are two main ways of looking at a living system: first, as a biological system with organization at several levels such as cells, tissues, organs, organs systems, and organisms; and second, as an information system (mind/consciousness complex) that parallels the physical one with pathways that provide control and regulation. Since consciousness has aspects that cannot be fully reduced to the brain, it must play a role in aging and in answering the question whether lifespan can be extended. In contrast, the aging of a nonliving system is determined by an increase in entropy alone for there is no information system that binds the structure together.[44]

Some aspects of the information organization belong to the conscious mind and thus are part of the cognitive system, but there are others that are autonomic, which operate unconsciously and regulate bodily functions such as the heart rate, its force of contraction, digestion, respiratory rate, pupillary response, urination, and sexual arousal, and the autonomic system is the primary mechanism in control of the

[43] Booth, L. N. et al. The aging epigenome. Molecular Cell 62, 728 – 744 (2016)

[44] Kak, S. Time, consciousness, and lifespan. Journal of Artificial Intelligence and Consciousness 12 (2025)

fight-or-flight response. It is likely that autonomic systems have an influence on aging, and there is evidence that the autonomic system can be retrained to a certain extent[45].

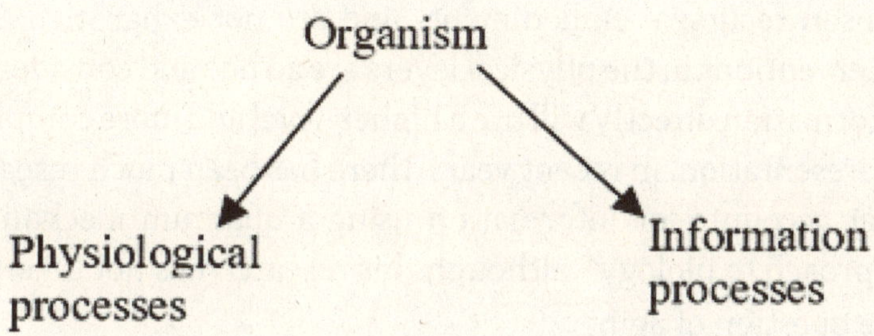

Unity of the organism mapped in two different ways

From the evidence in humans and mice that inherited deficiencies in DNA repair genes cause accelerated aging comes the theory that such damage is the primary cause of aging[46]. DNA damage affects transcription and cellular senescence so that a cell stops dividing permanently but does not die, and somatic mutations in the DNA can cause cell functional loss and disease. There is a correlation between nucleotide excision repair capacity and progression of aging and thus to extend the lifespan one needs to counter the disease processes that are caused by DNA damage.

---

[45] Streeter, C.C. et al. Effects of yoga on the autonomic nervous system, gamma-aminobutyric-acid, and allostasis in epilepsy, depression, and post-traumatic stress disorder. Medical Hypotheses 78, 571-579 (2012); Shobana, R. et al. Effect of long-term yoga training on autonomic function among the healthy adults. J. Family Med. Prim. Care 11, 3471-3475 (2022)

[46] Ou, H.L., Schumacher, B. DNA damage responses and p53 in the aging process. Blood 131 (5), 488–495 (2018); Vijg, J. From DNA damage to mutations: All roads lead to aging. Ageing Research Reviews 68, 101316 (2021)

The fact that the processes behind aging can be somewhat reversed by efforts at systemic, cellular, and molecular levels points to the possibility of even more dramatic results at the informational level since physical mechanisms only access chosen recursive embodiments and are not exhaustive. The interventions in the physical layers are ad hoc and considering information directly will use a higher-level and more complete representation. In recent years, there has been much research that accounts for information using a quantum mechanical approach to biology[47] although this research has not taken up the question of aging.

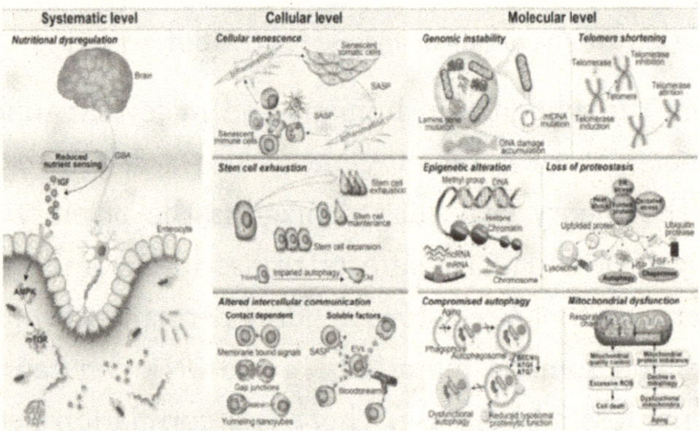

Ten hallmarks of aging in three categories[48]

[47] Brookes, J.C. Quantum effects in biology: golden rule in enzymes, olfaction, photosynthesis and magnetodetection. Proceedings of the Royal Society A: Mathematical, Physical and Engineering Sciences. 473 (2017); Goh, B.H., Tong, E.S., Pusparajah, P. Quantum Biology: Does quantum physics hold the key to revolutionizing medicine? Progress in Drug Discovery & Biomedical Science 3 (2020); Slocombe, L., Al-Khalili, J. S.; Sacchi, M. Quantum and classical effects in DNA point mutations: Watson-Crick tautomerism in AT and GC base pairs. Physical Chemistry Chemical Physics 23, 4141–4150 (2021)

[48] Guo, J., Huang, X., Dou, L. et al. Aging and aging-related diseases: from molecular mechanisms to interventions and treatments. Sig Transduct Target Ther 7, 391 (2022)

In the next sections, we first propose a formal representation of the aging process with the organism considered as an information system that runs in tandem with the physical one. We describe the quantum Zeno effect[49] that shows how quantum system evolution can not only be frozen, but it may also be steered in any desirable direction[50]. Then we show that the self-observation process provides the possibility of slowing the aging processes and thereby extending the lifespan.

Biological systems and information

In the view of the organism as processing information, one can consider it to be a general quantum system. The organism undergoes evolution, but it is also the observer who can slow down change; it is both the system being observed as well as the system that does the observation. The effect of the individual as mind (the observer) on the body is clear from the placebo and nocebo effects[51]; and the fact that in an extreme case an individual can die of psychological shock[52]. Thus the lifespan is not fixed by the biological systems alone since the parallel-running consciousness state has an agency and it can modify biological processes.

[49] Koshino, K. and Shimizu, A. Quantum Zeno Effect by General Measurements. Physics Report, 412 (2005)

[50] Misra, B. and Sudarshan, E.C.G. The Zeno's Paradox in Quantum Theory. J. Math. Phys 18, 756-763 (1977)

[51] Hall, K.T. Placebos. The MIT Press (2022) ; Kak, S. Time, consciousness, and lifespan. Journal of Artificial Intelligence and Consciousness 12 (2025)

[52] Laub, D. Thanatos and massive psychic trauma: the impact of the death instinct on knowing, remembering, avnd forgetting. Journal of the American Psychoanalytic Association 51(2), 433-64 (2003)

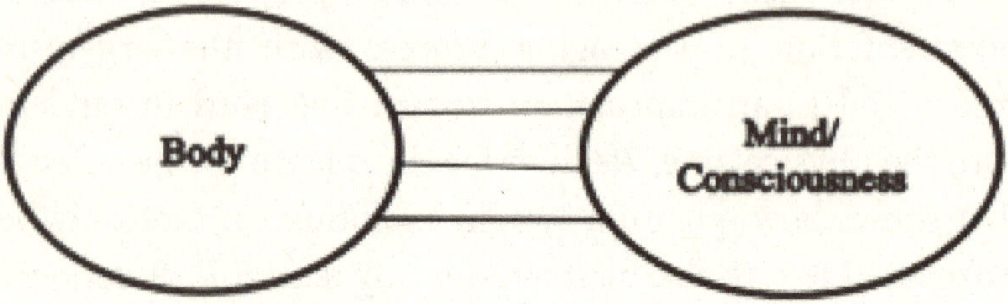

Body and mind as connected subsystems

It is important to note that the informational system is not completely mapped into biological networks because there are aspects to the observation process that transcend the body.[53] This is consistent with one of the interpretations of quantum theory that takes reality to be epistemic[54], which means that at the most fundamental level it is information. The above figure shows the human interconnected body and mind/consciousness subsystems.

The biological system and the informational system operate in tandem and at first level they can be viewed as being like the two sides of a coin, so that one cannot be derived from the other.

Quantum mechanics speaks of wave-particle duality, implying that in the wave aspect there is another variable that corresponds to the phase of the wave. The state of the system can be in a superposition of orthogonal states and the dynamics of the system are ordinarily governed by the unitary (reversible) evolution of the Schrödinger equation, together

[53] Kak, S. No-go theorems on machine consciousness. Journal of Artificial Intelligence and Consciousness 10, 237-247 (2023)

[54] Schrödinger, W. What Is Life? The Physical Aspect of the Living Cell. Cambridge University Press (1944); Penrose, R. The Emperor's New Mind: Concerning Computers, Minds and the Laws of Physics. Oxford University Press (1989)

with non-unitary (irreversible) transformations from the interactions with the environment. When a measurement is made the state function with its superposition states collapses to one of the components in a probabilistic manner.

If normal aging, with its corresponding lifespan, is associated with unitary evolution (mapped, say, into fundamental processes associated with the DNA), then the suspension of this evolution using the counter-intuitive quantum Zeno effect will provide a path for extending the lifespan.

The accounts of the capacity for extraordinary transformation through suitable practices based on mind-body unity are to be found in the documented power of yogis[55]. One can theorize that behind these abilities are methods that intuitively use a calming gaze on the mind. It may also be that some of these capacities arise from the access and control of the autonomic nervous system[56], which many yogic practices make possible.

The life state

The living state as a superposition of the life state and the death state is well recognized in psychology not only in cognition as mentioned in the previous section, but also in how one speaks of the simultaneous power of Eros (1) and Thanatos (0) that echoes Vedanta, in which one speaks of death as the dual (Yama in Sanskrit) to be the driver hidden in the unconscious (the causal mind). As mentioned before, the consideration of complementary variables as constituting the phenomenon of life implies a quantum mechanical description.

---

[55] Garbe, R. On the voluntary trance of Indian fakirs. The Monist 10 (4), 481–500 (1900); Murphy, M. The Future of the Body: Explorations Into the Further Evolution Of Human Nature. Penguin (1993)

[56] Kafatos, M.C. and Kak, S. Veiled nonlocality, cosmic censorship, and local observations. Physics Essays 28, 182-187 (2015)

The state may be proposed as an unconscious supervisory process sitting on top of other information mechanisms. It evolves slowly to the death state $|0\rangle$ and its evolution from $|1\rangle$ to $|0\rangle$ represents the system's lifespan.

Different measurement conditions in human cognition are complementary in the sense that they are mutually exclusive, and only one can be applied at any time; and they are both necessary for a complete accounting of the measurement. An important consequence of complementarity is that the sequence or order of the measurements matters.[57]

Measurement is a product of an interaction between the object of investigation (the mind) and the measurement instruments (the senses), and what is recorded is created by the measurement. For the energy-optimality of the cognitive system, beliefs, attitudes, and intentions do not exist in memory as properties ready to be recorded; instead, they are constructed when needed[58]. After the response to a question has been generated by the mind, this becomes the context for the construction of the response to a subsequent question, for it is still in the short-term memory. It is due to this that the probability p(AB) is not equal to p(BA) in a consciousness system, just as is the general case in quantum mechanics.

The simultaneous measurement of two complementary variables will also be associated with an uncertainty principle.

A quantum state operates in two modes: with the time-reversed unitary dynamics, and irreversible non-unitary observations or interactions, and the latter reflects the working of both conscious and unconscious mind.

---

[57] Grossberg, S. The complementary brain: unifying brain dynamics and modularity. Trends Cogn. Sci. 4, 233–246 (2000)

[58] Feldman, J. M., and Lynch, J. G. Self-generated validity and other effects of measurement on belief, attitude, intention, and behavior. J. Appl. Psychol. 73, 421–435 (1988)

Quantum Zeno effect

The unitary evolution of the quantum state may be suppressed by repeated observation, known as the quantum Zeno effect, an effect which has been demonstrated widely in the laboratory[59] as well as in theoretical studies.[60]

According to the quantum Zeno effect, the time evolution of any quantum system may be slowed by sufficiently frequent observations, or the decay rate of a physical system may be altered as desired.

Consider a physical system, say a qubit in the $|1\rangle$ state, which represents the excited state of an atom before it decays, and it evolves over time to the $|0\rangle$ state, which is the ground state of an atom after it has decayed. If we measure the qubit after letting it evolve from the $|1\rangle$ state in a very short time, then it will be projected to the $|1\rangle$ state with a high probability. If such a measurement is repeated many times at short time intervals, it will again have a high probability of being projected into the $|1\rangle$ state.

This means that if we continuously measure the qubit using projective measurements, we can effectively freeze it in the $|1\rangle$ state. The shorter we make the time interval between measurements, the higher the probability that the state will be frozen in $|1\rangle$. The effect also works for open system dynamics, time-dependent evolution equations, and general quantum operations in place of projective measurements, discussed in detail by Möbus and Wolf.[61]

---

59 Patil, Y. S., Chakram, S., Vengalattore, M. Measurement-induced localization of an ultracold lattice gas. Physical Review Letters 115, 140402 (2015)

60 Möbus, T. and Wolf, M.M. Quantum Zeno effect generalized. J. of Math. Phys. 60 (2019)

61 Möbus, T. and Wolf, *op cit.*

Repeated self-observation thus represents a way for suppressing evolution of superimposed function towards the death state $|0\rangle$. As observation can also steer the quantum state to a state that is counter to its natural evolution under the unitary transformation, it can reverse the unitary aspects of aging that are tied to the biological aging process.

Since the quantum state is running parallel to the other physiological processes within the body and is linked to them, the reversal of its directionality can affect the aging processes in a favorable manner.

Self-observation in yoga

Self-observation in yoga, called *svadhyaya* in Sanskrit, involves cultivating awareness of one's thoughts, emotions, physical sensations, and behaviors. It is the fourth *niyama* (ethical observance) in the eight limbs of yoga, and it translates to "self-study" or "self-reflection."

The goal of self-observation is to uncover the layers of conditioning, habits, and attachments that obscure one's true nature which leads to greater clarity, freedom, and spiritual growth.

Self-observation can rewire the brain by strengthening neural pathways associated with awareness and self-regulation. Research has shown that mindfulness practices increase gray matter density in the prefrontal cortex, which is linked to self-awareness.[62]

[62] Hölzel, B.K. et al. Mindfulness practice leads to increases in regional brain gray matter density. Psychiatry Res. 191, 36-43 (2011)

7

Managing Change

In the future advanced societies of humans, higher knowledge of the future will be sought by councils of thinkers beyond the confines of present-era countries and alliances of nations. It will be something like the United Nations, but different in its organization since the UN is in urgent need for reform. One idea is to create a Council of the Elders quite like the one imagined by von Braun for Mars that will consist of members representing different parts of society. The term council refers to a committee of people of wisdom, independent of age, they could be very young or more senior citizens of planet Earth.

Societal Challenges

The Council of the Elders will examine the clear and future danger of life being affected by disruptive astronomical and societal factors. We are arguing for such a council since the existing national and international organizations have not proven to be up to the task in dealing with the Covid pandemic or in stopping the Ukraine war. This Council will provide moral guidance to the existing political institutions.

Some suggest that the extinction of dinosaurs 66 million years ago was triggered by multiple causes: volcanism, marine regression, and extraterrestrial impact. In this scenario, terrestrial and marine communities were stressed by the changes in, and the loss of, habitats. Dinosaurs, as the largest vertebrates, were the first affected by environmental changes, and their diversity declined. At the same time, particulate materials from volcanism cooled and dried areas of the globe. Then an impact event occurred, causing collapses in photosynthesis-based food chains, both in the already-stressed terrestrial food chains and in the marine hierarchy.

Now science gives us tools to deal with future events like what caused the extinction of dinosaurs. The Council will take all living organisms that comprise life on Earth to be contained in the Earth Ark. It will look at life on planet Earth just as the legendary Noah's Ark addressed preserving life from the threat of the ancient flood.

Closely behind the challenge of diminished populations is the risk of wars. Local wars could morph into global conflicts with the use of nuclear weapons and millions of deaths. The third challenge is the possibility that technology will lead to unwelcome outcomes such as the dominance of AI in ways we cannot imagine. We take it for granted that the usage of AI will be benign but accompanying every bright outcome there is a dark counterpart.

There is a case to be made that nuclear energy for all its risks, if developed properly and safely, will be abundant, cheap and satisfy the needs of technological societies now and in the distant future. It will allow humanity to bypass the risks of global warming and everything associated with it.

A fourth major challenge is the change of the surface temperature of the Earth. As we contemplate global warming, it appears that the warming of the Earth's environment is already upon us and turning many parts of the Earth into drylands, with increased risks for fires, pollution from wildfires as well as risks to life, agricultural productivity and dangers to ecosystems. However, with the anticipated trajectory of a falling world population, this factor will not be as important as was feared in recent decades.

There is also the diminishing of fresh water, which implies that we will have to move to desalination processes of the abundant sea water. This will require large expenses and new technologies; but this risk will decrease as populations fall.

A fifth challenge is the state of health and risks associated with global epidemics and viral risks, as the COVID experience painfully showed us. The life-expectancy in many societies appears to have peaked. There is the problem of exploding rates of autism and various addictions amongst adults.

A sixth challenge is the state of technological, modern societies: Are we living better now than in the past? Is the quality and experience of life better? It must not be forgotten that Technology has increased disparity amongst social groups, and increased loneliness and depression.

A major risk to life on Earth is the possible close encounter with a supernova explosion. Such an event would affect all the planets and may end life on Earth. Even more numerous but not as devastating would be nova outbursts which are tied to the existence of double stars, where one of the stars ages and eventually leads to a nova outburst.

Homo Sapiens made its appearance on Earth some 250,000 years ago,[63] a mere 11% of time from the beginning of life on Earth, where all species evolved over many eons. Now there are two kinds of dangers that could wipe out humanity: first, is in the short term related to nuclear war, biological pathogens, or destructive genetic mutations; second, is in the long term with the destruction of Earth as a livable habitat for all forms of life.

For the short term risk of the destruction of human life on Earth, a colony on Mars or some another planet could then be used to reintroduce humans on Earth.

The long-term danger is from the inevitable evolution of the Sun, and in that case the situation for the Earth will dramatically change. When the Sun will be on its way to become a red giant, RG—its color will be more like orange red—the temperature of the Earth's surface will cause the oceans to boil away about 500 million years in the future, perhaps much sooner than that, maybe 300 million years and long before the full development of the RG phase. Life took about 4.5 billion years to reach its current state, and 300 – 500 million years may seem very long into the future, however this doesn't imply that this danger should be

[63] Baum D.A. The origin and early evolution of life in chemical composition space. J. Theor. Biol. 456, 295–304 (2018)

ignored. It will come to pass and this will be the end of all life on Earth.

The future expanding RG Sun as it may look
(photo of sunset by Menas Kafatos)

This major challenge will dwarf all other risks arising from ongoing changes in the solar system, comet and meteor strikes, the motions of the continents in the ancient continental drift and everything else that occurred in the past or will likely occur in the future, including pollution, global warming and now the march towards underpopulation of humans. This is why the mega risks associated with the evolving Sun remain a stark reality that future humans cannot under any circumstance ignore.

This ultimate risk will be a game changer affecting every life form, micro-organisms, ancient cellular life that still populates the depths of the oceans and on the ground, multicellular organisms, plants, trees, everything in the living oceans, animals on land and in the air, various ecosystems, that is all life.

Everything eventually comes to an end and so will all life, unless we collectively decide to know the risk and do something about it. There will be a global challenge not seen before. It is not science fiction, it is as real as the Sun shines, and will keep shining, eventually becoming brighter and brighter and threatening all life.

Spaceship Earth

As the Sun begins the pathway to the red giant phase, conditions will become unbearable for life to exist on Earth. The planet will become a crispy rock, and humans must address the developing situation well in advance. Humanity will basically have four choices over millions of years:

The first choice would be to stay on planet Earth and move the planet "out", where it will be safe, such as in the vicinity of Saturn and then when the Sun becomes a white dwarf, to move "back in" close to the source of light.

The second choice would be to populate Ganymede or further out Titan, but, as the Sun eventually becomes a white dwarf, with too little light, we will have to move in again, closer to the WD.

The third choice would be to form large space colonies and travel with these colonies to the stars. This will be a repeat of what the Europeans did when the horizons of their land became too small and they sought new lands beyond the seas, or as the Polynesians did in their heroic path to populate new islands in the Pacific. Each space colony would be a complete ecosystem.

The fourth choice may be termed *science fiction* by our current knowledge. It would be to move the entire Earth into a favorable location around another star. In a sense spaceship Earth will jump through space-time into the neighborhood near another star, in sort of "island hopping."[64] Of course, this is an impossible choice for it contradicts the principles of physics as we know them.

The first and second choice involves remaining in the solar system and addressing the changing conditions of the Sun's expanding radius and rise in luminous output. While the third choice would be to move out of the solar system into the galaxy, beyond the solar system. This choice of, traveling to the stars in spaceships will realize Noah's Ark in the distant future. Humans would become extraterrestrials, reaching out to distant planets in the galaxy.

If the laws of physics turn out to be radically different from how we understand them when general relativity and quantum mechanics become unified, new possibilities may materialize. Behind this speculation is the possibility that general relativity wormholes could move Earth away from where it is located now.

Harnessing nuclear energy and the forces of Nature, including gravity, in unknown ways will open new worlds for exploration. The so-called quantum vacuum is anything but empty, it contains vast amounts of energy in latent form. The task would be to access the plentiful energy of empty space and use it for new ways to sustain life.

---

[64] Carroll, S. The Biggest Ideas in the Universe: Space, Time, and Motion. Dutton (2022)

All options will be examined to utilize the laws of physics, and create new technologies. The tasks we will be facing have surely been faced by countless civilizations throughout the universe.

One may speculate on the possibility of transporting spaceship Earth with all its ecosystems, and all living species into other regions of space-time, requiring tunneling in spacetime. These tunnels are the wormholes and they have been postulated to exist at microscales. They may also exist at macroscales, linking regions of spacetime allowed by General Relativity.

For such tunneling to become a reality, three conditions must be satisfied: First, macroscopic wormholes exist as GR allows. Second, entanglement of all particles of the Earth into a coherent whole at an instant of time will have to be a reality. The third requirement is the existence of non-locality at macroscale. Researchers believe that nonlocality is a universal property of the world, regardless of how and at what speed quantum particles move.[65] But the entanglement of all particles appears to be impossible.

Exoplanets in several regions in nearby stars can support life. Designed to survey a portion of Earth's region of the Milky Way to discover Earth-size exoplanets in or near habitable zones and to estimate how many of the billions of stars in the Milky Way have such planets, the Kepler space telescope, observed 530,506 stars and detected 2,778 confirmed planets as of June 16, 2023. Located at different

---

[65] Streiter, L.F., Giacomini, F., Brukner, Č. Relativistic Bell Test within Quantum Reference Frames. Phys. Rev. Lett. (2021)

distances from the Sun, the nearest being Proxima Centauri about 4 light years, to beyond 1,000 light years, they tend to be like Earth, however quite often a bit larger.

A moral dilemma that arises is that unless we put all living species into many space colonies, we will have to make hard choices and leave many of them behind. However, further research advances may make it possible to create specific species conveniently at remote locations based on their genetic code. These advances will also make it possible to revive species that are extinct.

Project Omega will involve many decisions that are presently beyond our consideration. As of now, we have no framework to consider, no prior experience and even no moral basis to make such decisions. We will require a radical new way to think collectively, beyond traditional ways and consider our species as caretakers of all in the Earth ecosystem. To fulfill such a mission, we will need our progeny to think and act globally.

The Future of the solar system

Earth and other planets continue to be bombarded by meteoroids, but significant impacts are rare. For instance, a large impact like the one that caused the extinction of the dinosaurs is expected roughly every 100 million years. Relative to the lifetime of human civilizations, this would seem to be a rare event, however over five billion years which is the current lifetime of the Solar System, it would not be rare.

Over thousands of years, the orbits of planets can slightly change due to gravitational interactions. For instance, Mercury's orbit is slowly expanding due to tidal forces from the Sun. Mars might see changes in its ice caps due to axial tilts, while Venus could experience shifts in its incredibly dense atmosphere, though these changes are minimal on a human timescale.[66]

The Sun is currently in the main sequence phase of its life. It will eventually become a red giant, expanding to potentially engulf Mercury, Venus, and possibly Earth. After this phase, it will shed its outer layers and become a White Dwarf. In this process, atmospheres of planets like Earth could be lost or transformed. Earth's atmosphere might be stripped away by solar wind as the Sun's activity increases. More ominously, the oceans will begin to boil away way before the Sun becomes a fully developed red giant.[67]

The Moon is gradually moving away from Earth due to tidal interactions. This process will continue, affecting Earth's rotation and tidal patterns. It is interesting to speculate how the combined Earth - Moon would evolve and how the evolution of the system will affect future living conditions.

There's a small chance of giant impacts or close encounters with other celestial bodies that could drastically alter planetary environments or even knock planets out of their orbits. These would clearly be major challenges for life to survive.

The solar system orbits the Milky Way galaxy, and this will in the long-term future bring Earth into

---

[66] Garvin J.B.et al. Revealing the Mysteries of Venus: The DAVINCI Mission. Planet. Sci. J. 3, 117 (2022)

[67] Harvey, R. The Solar System: Exploring the Sun, Planets and Their Moons. Amber Books (2022)

different environments, potentially affecting cosmic ray exposure or gravitational interactions with other stars. As mentioned before, as the Sun moves into regions with young massive stars, a major risk might prove to arise from future supernova explosions.

After the Sun becomes a white dwarf (WD), the planets will continue orbiting this stellar remnant. The white dwarf Sun will eventually cool to become a black dwarf.

When the Sun balloons up into a red giant, it will consume Mercury, probably Venus, and maybe even Earth. Even if the outer planets avoid being swallowed up by the expanding Sun, they might eventually get pulled in or ejected from the Solar System. Recent discoveries suggest they can survive intact.

Using NASA's JWST space telescope, astronomers have for the first time directly imaged planets orbiting around white dwarfs. The planets follow orbits resembling those of the giant planets in the outer Solar System—big enough for them to have escaped the inferno.

A star's red giant stage ends when it expels its outer layers and shrinks into an Earth-size white dwarf. Astronomers have already seen hints of planets surviving around these faintly glowing cinders. For example, the atmospheres of some white dwarfs appear polluted with rocky material, suggesting unseen planets are constantly deflecting comets and asteroids into them. Other white dwarfs periodically dim, suggesting they are eclipsed by orbiting planets. Potential planets orbiting a white dwarf have been detected.[68]

White dwarfs are good places to look for planets because they shine with just 1% of the brightness of the Sun. That makes it easier for telescopes to block out the star's light to reveal the faint glow of surrounding planets.

Around some white dwarfs, objects that appear to be planets have been spotted. If it is confirmed that these are planets, it would suggest there is an abundance of such planets available for JWST to observe. Because such planets are bright compared with their stars, their atmospheres should be relatively easy to study, to discern their similarity to or difference from the Solar System's giant planets.

The intriguing possibility arises that even stars that have reached the end stage of low mass evolution as white dwarfs, could still support the existence of planets around them.

The situation is of course different for massive star evolution, which are several solar masses, because such massive stars eventually explode as supernovae. Even if planets exist around such stars, the rapid evolution of the parent star would doom any life that was developing to reach mature stages, and the maelstrom of the supernova explosion would end everything. Around such massive stars. life could never develop.

A large percentage of stars are double or multiple star systems. Planets orbiting such stars, either one star or

[68] Limbach, M.A. et al. The MIRI Exoplanets Orbiting White Dwarfs (MEOW) Survey: Mid-infrared Excess Reveals a Giant Planet Candidate around a Nearby White Dwarf. ApJL 973 L11 (2024)

both, would face conditions that are as challenging as the survivability around a supernova. Although life may have a chance to develop, the orbits of the two stars it may be a more difficult process but we don't know.

As far as we know there is only one star system, our own, where life successfully developed. We must understand not just the challenges and intricacies of biological evolution here on Earth but also understand the impact of astronomical factors. It is estimated that 85% of stars are double or multiple star systems. If we count Jupiter, even our solar system was almost a double star. To have a stable single star system where life can develop and flourish in one planet, Earth, not counting Mars, over billions of years may be a cosmic luxury.

An extreme point of view is that we are alone in the universe. It is hard to imagine this would be the case. If it is so, one would then ask, why did life develop here on Earth? What is particularly unique about Earth in the entire cosmos, the galaxy, and the universe? We do not subscribe to such an extreme view, but we really don't know. Our uniqueness remains an open issue.

Moving out and moving in

The choice for both the short term and long term risks is to move out and then move in. Moving out in the short term is to plant colonies on Mars and other moons in the solar system and return when the danger has passed, and for the longer term is to move out to the outer solar system, to escape the boiling away of the oceans as the Sun begins to expand in size and begins to become more luminous and hundreds of times or more than the current Sun.

The planetary inner realm's doomsday will bring a brief new dawn to the outer solar system. Saturn's moon Titan, the only known satellite with a significant atmosphere, would keep its dense shroud. It could enjoy several hundred million years of potentially habitable conditions with liquid oceans of water-ammonia, simulations show. The environment may even resemble early Earth.

Eventually, Pluto and its cousins in the Kuiper Belt — plus Neptune's moon Triton — could be the most valuable real estate in the solar system. When the Sun turns into a red giant, the temperatures there will be equivalent to those in tropical locales on Earth today.

When our Sun approaches its red giant phase, it will run out of fuel in its core. As hydrogen fusion slows, the core begins to contract. As the core gets smaller, it heats up until it can kick off another round of nuclear reactions, fusing helium through triple alpha reactions, first into carbon and subsequently into heavier elements such as nitrogen and oxygen. The hotter core also makes hydrogen fuse in the "shell" of material surrounding the core.

Meanwhile, extra heat produced deep within the star causes its outer layer of gas to puff up, producing spectacular gaseous planetary nebulae (having nothing to do with "planets") and enriching the interstellar medium with heavier elements. All these processes are not even considering the much more powerful supernova explosions of massive stars, producing heavier elements up to iron through fusion and beyond iron through the s-process, eventually producing Uranium, other super heavy elements and which result in neutron stars or black holes in the final, end stages of stellar evolution.

Returning to the end red giant phase of stars of lower mass like the Sun, is typically the most violent time in a star's life, producing stellar cores up to the Chandrasekhar limit of for stars up to approximately 1.4 solar masses. The bloated, dying star ejects material from its outer layers in intense episodic bursts, eventually leaving behind a collapsed stellar core, the White Dwarf, that is expected to outlast even the universe itself. In our own solar system, the Sun will puff up so much material that it will melt, evaporate and eat up some of the inner rocky planets and their satellites.

Scientists studying scenarios whether Earth will be both toasted and swallowed create computer models of the gravitational interactions between Earth and the Sun as the solar system ages. Recent modeling suggests Earth probably won't make it out intact. But even if it does, all the planets that survive will experience skyrocketing levels of energy from the expanding Sun, bathing their atmospheres and surfaces with intense radiation. Only the outer solar system planets, gaseous in nature and their multitude of satellites will survive the evolving Sun.

The planet orbits may also become unstable. As the red giant loses mass, the star's gravitational hold on its planets becomes much weaker, so their orbits will expand. In our solar system, the Sun will throw off about half its mass, so the outer planets' orbits will drift outward, settling twice as far in distance as they are today. Neptune, at 30 times the Earth-Sun distance, will eventually hike out to about 60 times the Earth-Sun distance.

Approaching the end of its hydrogen fuel-burning life, the Sun will become much brighter and bigger. It will grow so large in diameter that it could fill the entire sky from the surface of a surviving planet.

Life requires liquid water, and a rocky planet needs a certain amount of starlight so that any available water won't evaporate away or freeze. If scientists know enough about a star and its planetary system, they can calculate whether a particular planet could indeed harbor liquid water, meaning the star would be in the "habitable zone", allowing biology to operate and life to evolve.

But habitability of planets changes as a star gets older. Frozen planets, too cold to host liquid water when their star resembles our present-day Sun, may thaw out when their star expands into a red giant, because they will be receiving much more heat energy than before.

As result of the future evolution of the Sun, the choices facing our progeny will be numerous. Science will provide tools of understanding as well as the means to carry out more rapid evolution at all levels, biological, societal and individual.

The function and understanding of science will change as it molds our physical and social environments and, in turn, gets affected by them, which will create a true global civilization.

Humanity will consider what is optimal for humans and what is best for life in the future Earth, so that it can flourish and collectively survive.

It is here where the visionary possibilities of the future can be combined with the visionary possibilities which always existed and can be traced back to the great philosophies of ancient Hellas (Greece), from the pre-Socratic, to Pythagoras, Socrates, Plato, Aristotle, the neo-Platonists and a host of other inspiring systems, influencing the entire ancient world, Rome and the collective West; to the great achievements of Indian non-dual traditions, particularly Advaita Vedanta, Triadic Shaivism; as well as entire systems of ancient Eastern traditions, Confucianism, Taoism and Buddhism.

These ancient jewels of wisdom not only produced the best in philosophical thinking, but many of them were also at the core of the development of science. Coupled to benign religious systems, they helped propel humanity to the Renaissance, non-duality and collective advanced knowledge of the physical and biological sciences.

Our hope is that with these solid foundations humans will continue to expand, flourish and provide paths of meaningful choice for the collective ecosystem of living Earth.

Moving to Safe Spaces

As the Sun sheds its outer regions and eventually collapses to a white dwarf (WD), living near Titan or on Titan itself would not be a desirable future outcome. Life on Earth would face a stark frozen reality, the future WD being nothing but a speck of white light in the vastness of dark sky. It would surely be time to "move in", from the outer to the inner solar system.

As just discussed, after the RG phase, the protected Earth will have to somehow get close to the WD as the WD luminosity will be very low compared to what is now living around the Sun.

The WD will evolve extremely slowly, in fact outlasting even the universe. Eventually the WD will cool down and it will then become a black dwarf. However, such a phase will be so far into the future that it is of no concern to present humanity.

The stability of the WD will allow our progeny to live for extremely long times, not worrying about the challenges of stellar evolution that the Sun underwent billions of years in the past from the WD state. Although the future dwarf Sun will be very dim by today's Sun standards, it will provide a long-term stability, with constant luminosity spanning many eons. There will be plenty of ages for collective life to survive, evolve and consider all possible paths of evolutionary developments.

This viable environment will last for what would seem to be an eternity, allowing humans and all future life to remain stable and eventually non-evolving in the biological sense. Evolution will be guided for the entire ecosystem of living Earth, and it may allow for possibilities that we cannot even fathom, allowing for mind to expand, and for the universe to be comprehended in ways not done before.

It is worth contemplating what our progeny will be undertaking in the distant future. Here in brief is a non-exhaustive list:

- From the side of the Earth not facing the WD, they will be studying the dark skies (it is not clear if the Earth will still possess a Moon, most likely not), and they will be studying molecular clouds, the hubs of future stellar births and associated planetary systems.
- They will be studying producers of high energy particles and radiation, emanations from pulsars, accretion onto black holes and of course massive active galactic nuclei including the center of our own Milky Way Galaxy.
- They will be searching for signs of ET signals and perhaps they will be making contacts with distant civilizations and exploring the possibilities of interstellar travel and reaching nearby planetary systems.
- Having all the time in the world and vast "libraries" of data, they will be studying ancient scientific and philosophical systems, the advent of new physics and

new biology, the wisdom of antiquity, new advanced mathematics and, possibly, the replacement of languages and linear interpersonal communications with direct mind-to-mind contact.

- Such a world of expansive thinking will allow humans to reach the source of wisdom and benevolence which the ancient thinkers believed and taught is within each one of us.

8

Ancient Wisdom, Cosmic Consciousness

In the naïve view, the world is an aggregate of physical objects and related forces, but this view has been shown to be wrong by quantum theory in which consciousness is complementary to materiality, and despite considerable efforts scientists have been unable to bring the observer within the framework of the theory. As shown in earlier chapters, there are several arguments that show that consciousness is not computable, which means that it is not a property of matter.

Some theories assume that consciousness is a property arising out of the complexity of the brain's neuronal interconnections. But a material property cannot develop into self-awareness, which is clear from the fact that hardware may be seen as a mapping of software. Awareness or consciousness cannot be physical, although our minds and their contents are physical.

Much confusion in the academic approach to consciousness arises from its conflation with mind. The correct view is to assume that the mind is the apparatus used on the neural ground by which observations are made and experienced. If consciousness is a non-material category, then to speak of its evolution makes no sense. Certainly, there has been an evolution of the mind, and we can see a spectrum of minds across animals. One also sees a spectrum of capacities when a person wakes up from anesthesia.

It is a commonsensical premise of science that the physical world (which includes the biological world) is causally closed. Causal closure implies that all physical events must be due to physical causes leading to the question as to the source of the freedom cognitive agents appear to possess. So how to explain the origin of observers? And if observers are not zombies in a mathematical universe as has been suggested, one needs to explain the apparent influence of observers on physical systems in physics and neuroscience. Observers and agents deal with information, therefore, it is also important for understanding the nature of physical information, whether agency could be achieved in artificial intelligence machines, and if consciousness is computable.

The "Cambridge Declaration on Consciousness in Non-Human Animals" that was mentioned in an earlier chapter states that the capacity for consciousness likely emerged very early in evolutionary terms, and those processes that support consciousness in humans are likely characteristic of many living creatures. In fact, according to the declaration, based on several considerations from comparative brain anatomy and current knowledge about the neurobiology of consciousness, it is almost certain that some form of consciousness is present in all mammals and could have emerged on the evolutionary timeline at the branch point of amniotes. This conclusion is consistent with the recent Cambridge declaration that occurred on July 7, 2012, at the first annual Francis Crick memorial conference on consciousness.

"The absence of a neocortex does not appear to preclude an organism from experiencing affective states. Convergent evidence indicates that non-human animals have the neuroanatomical, neurochemical, and neurophysiological substrates of conscious states along with the capacity to exhibit intentional behaviors. Consequently, the weight of evidence indicates that humans are not unique in possessing the neurological substrates that generate consciousness. Non-human animals, including all mammals and birds, and many other creatures, including octopuses, also possess these neurological substrates."

Main propositions

According to the commonsensical view of reality, the world consists of nonliving, insentient matter and living things. The division between the sentient and the insentient is of vital importance to the foundations of science. To the extent, sentient beings are conscious, they have free will and therefore their future states cannot be fully predicted. However, since sentient beings also have physical bodies, the outcome even if free will be within the constraints of the physical system and so their behavior can be mapped probabilistically.

An insentient object can be fully described, assuming that a complete theory for it is available, and given its description and the current state, its future states can be estimated. Science is about describing insentient objects and predicting their future states and interaction with other objects.

We will use the term "observer" for sentient object irrespective of their size since such an object is "aware" of its environment and takes that into consideration in its future action. This active engagement with the environment makes the task of defining the size of the observer problematic. In reality, this is an issue in the definition of an insentient object as well for it is in continual short-range and long-range interactions with other objects.

Proposition 1. The universe consists of insentient objects and observers.

Proposition 2. Different observers use the same ground consciousness. Each observer is simply a different mind, which is the instrument of observation, and the same observer can also be in entirely different mind-states.

Proposition 3. The capacity to observe depends on the nature of the brain structure and how much of it has been recruited for the observation.

Proposition 4. A cloak of probability covers the observation process. There is no way to tell whether the observer made a free decision or whether it was a random event.

Mind body medicine

Allopathic medicine is based on treatment using pharmacologically active agents or interventions to treat or suppress symptoms or pathophysiologic processes of diseases or conditions. It normally looks only at specific conditions and uses protocols that are determined for large populations. There is little attention paid to how the mental state of the patient could affect the treatment of the disease. This is relevant as the brain states and the immune system are linked.

Until recently, it was assumed that the central nervous system (CNS) is cordoned off from the peripheral immune system and is reliant only on its resident immune cells called microglia. Peripheral immune cell breaches anywhere in the CNS were considered signs of disease. But researchers now know that diverse immune cells — possibly by the millions — circulate in the cerebral spinal fluid (CSF) and live in the brain's outer membranes even in healthy individuals.

These factors are not accounted for in studies on efficacy of drugs and, therefore, the test and control samples may not be robust enough to draw the published inferences.

The allopathic paradigm itself is being questioned, and people are asking if the body (together with the mind) needs to be viewed in a holistic manner as an ecological system.

There is a deeper problem to which little thought has been given. This is that binary logic, which is at the basis of experiment design, is not optimal. Ternary logic is optimal, and it requires the hypothesis to test between three classes that may be called "true", "false" and "maybe".

We are not speaking of multi-class classification which is common in data analysis; rather, we are speaking of an underlying triple logic associated with the fundamentals of the problem, which, to the best of our knowledge, has not been examined before. It is significant that a ternary classification of patients is used in at least one medicine system (Ayurveda) as a balance between three doshas. Furthermore, the activity at a more basic level is also seen as a balance between the three characteristics of sattva, rajas, and tamas.

Evolution of mind

If consciousness evolved in conjunction with cephalad development of the central nervous system, then its emergence should, in principle, be identifiable at a discrete point on the tree of evolution.

To identify the origin of sentience along an evolutionary timeline, it is beneficial to consider a common element that might link consciousness across species, rather than focusing on the ostensibly unique qualities of human experience such as self-reflection. Furthermore, this common element should likely relate to a goal-directed behavior or response pattern that confers a survival advantage in each environment. Consider, for example, the unicellular paramecium, which is covered with several thousand cilia. These cilia can serve both the function of sensing environmental stimuli and initiating motility responses (e.g., attraction, avoidance) based on the nature of those stimuli. This preneural example of a single structure (i.e., cilia and their coordinated activity) mediating both sensation and response is intriguing but does not establish the primacy of motility as a kernel of consciousness.

To identify which aspects of mental machinery should be the focus of evolutionary consideration for consciousness, we need to identify the neural correlates of the most primitive core of human consciousness. However, studies of the content of consciousness (e.g., the awareness of a red rose placed in your visual field) already assume a conscious brain; thus, the neural activity or structure identified in these paradigms corresponds to a specific content within a preexisting consciousness. Studying the level of consciousness (e.g., arousal states) is also beset with difficulties.

Traditionally, the emergence of consciousness (as judged by the return of a response to command) was correlated primarily with activity of the brainstem (locus coeruleus), hypothalamus, thalamus, and anterior cingulate (medial prefrontal area). Surprisingly, there was limited neocortical involvement that correlated with this primitive form of consciousness.

Another approach that is used is that of taking consciousness as a property of different kinds of brain networks. This makes it possible to test hypotheses across species, where functionally similar cognitive systems may arise from neurobiologically distinct structures. For example, the mammalian cortex and avian pallium look different anatomically when examined under a microscope, but they may support similar network functions that can be quantitatively assessed and compared with results known for the human brain.

From a cognitive perspective, the meaning of self-awareness behaviors in a mirror remains somewhat controversial. Some argue that mirror behavior could be more easily explained by simple knowledge of one's body. The neurobiology of having a body sense is something that is highly linked with a sense of consciousness. It has been suggested that all an organism requires to self-recognize is a mental representation of its own physical self; the organism matches the kinesthetic representation of the body with the image seen in the mirror and recognizes itself.

We have noted that the brainstem, diencephalon, and limited association cortex capable of recurrent processing is consistent with a core or primitive consciousness. However, what accounts for the richness of human experience in contrast to those of early mammals or birds? Being a subjective experience, the level or quality of consciousness across species cannot be compared. Although humans can do poetry and abstract reasoning, it is difficult to imagine that a sedentary human has a higher level of consciousness than a highly alert animal fleeing a pursuing predator.

Despite these considerations, human consciousness—especially the capacity for self-consciousness and reflection/projection in time—seems unique. Evidence suggests that the core of consciousness is rooted in phylogenetically older structures such as the brainstem and diencephalon,[69] the evolution of that which is particular to human consciousness may be more closely associated with the development of the

[69] Merker B. Consciousness without a cerebral cortex: a challenge for neuroscience and medicine. Behav Brain Sci. 30(1), 63-81 (2007)

frontal cortex. The relative size of the frontal lobes with respect to the total neocortex is roughly the same in modern humans and great apes, but richer interconnectivity might account for advanced cognition in Homo Sapiens.

This information flow reversal suggests that human consciousness is more defined by internal dynamics than external stimuli. It may help explain, why certain animals arecapable of an MSR response. According to one theory, human consciousness is a closed system or "oneiric" (dreamlike) state that is modulated by environmental input, a theory consistent with REM sleep as a building block for human consciousness. The relative independence from environmental determination of conscious content would potentially permit a greater diversity or richness of experience in comparison with species without dominance of anterior-to-posterior flow. This independence would also facilitate the projection and simulation associated with plans, which is of clear relevance to survival. But the role of information flow in consciousness is unclear currently and requires further neuroscientific investigation.

All evidence points to mind being a product of specific structures in the brainstem and diencephalon in conjunction with neocortical connectivity and recurrent processing can result in primitive minds that are like that of the human. Finding these structures in mammals and birds we can assert that their minds have similarity to ours.

This argument reveals that one cannot leave out the possibility that there may be further evolution in the biological realm that goes beyond what has been conceived in science.

The outer and the inner

The old divide between science and religion is counter productive and unhelpful for a deeper understanding of reality. In the grand sense, philosophy is the parent of both religion and science, the two human endeavors that in the past seemed counter to each other. In fact, they are not mutually opposed because they both wish to understand the mystery of our existence.

The Indian or Eastern approach to these questions is more philosophical and metaphysical because of the centrality of consciousness in it, whereas the Western approach is philosophical with a focus on objective classification. This is, of course, a somewhat general division and does not necessarily correspond to integration of basic and higher principles that both the East and the West have been seeking for thousands of years.

The ancient philosophical systems of India, Hellas, and China achieved a profound understanding of knowledge in the service of the collective whole, each in a unique manner. There was knowledge of physical, mental, and higher principles, and narratives on the experience of the qualities of existence, the qualia. We now have mathematical ways to formalize qualia and the relationships to consciousness, namely in terms of sheaf mathematical formalism.[70]

The global understanding being sought over the centuries was brought about by not one person or one tradition, it

[70] Kafatos, M.C. and Kato, G.C. Sheaf theoretic formulation for consciousness and qualia and relationship to the idealism of non-dual philosophies. Progress in Biophysics and Molecular Biology131, 242-250 (2017)

has been a collective understanding of philosophers and thinkers.

In ancient India, the Vedas describe the mystery of existence as a unitary reality that is experienced in many ways. Later philosophical systems (Darśanas, "ways of seeing") put it down into six major perspectives of ritual action (Mīmāṃsā), logic (Nyāya), categories of reality (Sānkhya), physics and observers (Vaiśeṣika), synthesis of experience (Yoga), and cosmology (Vedanta). To this were in time added several new philosophies such as that of grammar (as a way of understanding the world through language), Shaivism, Vaishnavism, Buddhism, Jainism, and so on.

These schools later spread to many lands of Asia, and Triadic Kashmir Shaivism came to be a particularly insightful approach to consciousness. In ancient India, philosophy was not just thinking of the highest order, the teachings were meant to liberate the seeker from the confines of the physical body and the limitations of the mind.

In ancient Hellas, we can follow the wisdom of the great philosophers, and their collective views transcending boundaries of space and time, and the principles operating underneath appearances. In ancient China, traditional teachings sought different ways of right living and understanding.

From our personal experience, we believe it helps to read the teachings in the original language. Thus, one gets deeper appreciation of the material upon reading Plato's Timaeus, Politeia (City State), and Kritias in ancient Greek, and not modern Greek or English or other languages. Aristotle's Nichomacheian Ethics also provides profound understanding of what ethical life is.

The need to read the ancients texts in Sanskrit holds for the multitude of Indian works, the Vedas, Advaita Vedanta, the Sutras of Shaivism, the rich literature of Tantras which promoted higher thinking even in the everyday processes, as non-ending processes of emanation, sustenance and re-absorption. Of course, good translations are also needed, but language is an important defining aspect of philosophical treatises as taught by the great Shaivite philosopher Abhinavagupta.

The deep works of different traditions seek unity. This unity amid differences and variations is also the goal of science. Therefore, the aims of philosophy and science in the end are similar; they point to unified understanding and integration beyond the kaleidoscope of differences.

In philosophy, spirituality and science, language plays a fundamental role. In science, the language is mathematics, including number theory, geometry, algebraic geometry and topology. In spirituality, one needs to be fluent in the language in which the experience has been communicated. To obtain a true understanding of the Bhagavad Gita, one needs to read the 700 verses in Sanskrit. Likewise, in Buddhism, the text of the Diamond Sutra should be read in Sanskrit.

This does not imply that good translations are useless, far from that. It is just to keep in mind that subtleties in the original language may be missed. The sage Abhinavagupta knew about the importance of right interpretation and as an expert in language and grammar advised his disciples to undertake the correct reading of the ancient texts.

There is a vast literature and teachings of the religious founders who sought Union with the One, going beyond specific traditions, historical developments, intercomparison between spiritual paths and apparent differences.

One must accept that that the philosophy of ancient wisdom traditions and modern science, particularly quantum mechanics and modern quantum field theory, are related. The word yoga in the full sense of the term means union with the universe, that is union with non-dual universal consciousness. Ultimately, we must tap the collective Wisdom of the Ages, irrespective of origins, irrespective of the local conditions.

Revisiting wisdom

If we aim to comprehend the world, we must first proceed from an understanding of the role of the mind. Logic and mathematics are at the core of science in creating the discernment of the collective units that together make up the world. When we use the term "understanding", we do not mean or imply mere intellectual comprehension. What we mean by it is what the "stuff" of the mind really is.

The mind is a challenge, for it is not a physical object. It deals with objects and things beyond them, as well as with the qualia of experience. It is here where quantum mechanics becomes not only useful; it is necessary to understand the role of the subject, the object and all processes linking the two. In fact, the founders of quantum mechanics held the view that it was basically a science of the mind, not of inert physics.

Reality is one, however interpretations of aspects of it are numerous. The uncertainty principle of quantum mechanics, when interpreted philosophically, allows for the many possibilities that exist to become one specific manifestation based on the choices the observer makes. The term often used is "collapse of the wave function" or the more scientific expression "reduction of the state vector" indicating that interaction with the physical world reveals an aspect of it to the observer. Instead of the mechanistic, rigid classical world, the world of the quantum is exquisitely dynamic, responds to the observer and as the uncertainty principle holds sway, it allows for freedom of choice, freedom of meeting different aspects of the one Reality in a kaleidoscope of possibilities and adding to these aspects in non-linear ways, in richness beyond imagination.

We will surely travel beyond the limits that our own minds put up against ourselves and everything around us. If one were to adhere to the belief of duality, it leads to a fragmented world view, where cooperation is difficult and might even be undesirable or impossible. However, the principle of complementarity of quantum theory allows both diversity and unity of the opposites.

It is time for humans to face our true collective mission while respecting and preserving traditions and the wisdom of the past and, even more importantly, base life on true moral and ethical principles. Studying the meaning of societal structures and transformations must also be done with an understanding of the past for that will be relevant to future human societies.

Until we wake up from the impediments created by the limited ego and the slumber of the narrow mind, we will continue stumbling and may even self-destroy and take with us countless, innocent living forms, as well as our young babies and children. This would surely be the mark of foolish beings, who don't understand their place in the Cosmos, who defy higher purpose and the wisdom of the ages.

For our progeny to map out a future, they will need to consider the sum of evolution of everything on Earth. Without being able to map the future, this would be impossible in the quantum view of the cosmos. We and our progeny welcome change as the everlasting dance of cosmic consciousness. Vast possibilities exist everywhere in the universe, including in our solar system and here on Earth.

In the quantum view of reality, existence is coupled to observation and the understanding that we observers form one observer, the One. It views all objects of existence as a vast tapestry of interactions extending over the infinity of time.

Over all dimensions of space, from the microscales of quanta to the vast extensions of the universe and to vast numbers of other universes, existence is primary, and changes are ever happening. The vision will ultimately bring together the microcosm and the macrocosm, in the long-sought unification of the physicists. This will include evolution of all life here on Earth and everywhere in planetary systems.

Evolution is occurring at all scales, over all objects and all the time. What brings everything together is observation and that is how consciousness influences it. At the mundane level, observation leads to formulating hypotheses, which leads to understanding, and to wisdom.

Wisdom is a collective undertaking, it is what unites science and, beyond science, brings together the mundane and the spiritual.

New ideas for physics

Physical reality is experienced through the interplay of time, space and memory. Of these, we know time by the change as in the succession of day and night and by the many processes within and outside the body. The sense of space is related to the three-dimensionality of the rooms that people spend time in; when flying high above the clouds, with the horizon seen more clearly, space doesn't appear rectangular.

If space were framed by three orthogonal axes that are at right angles to each other, there should be evidence of it at the cosmic scale. Specifically, spiral galaxies should have flat disks. But the map of the Milky Way galaxy, painstakingly constructed using information on pulsing stars called cepheids, shows that it is warped, and not flat.

It is our intuition that guides our sense of physical space, and the three-dimensionality of space is a convention. The unstated assumptions underlying our praxis become a habit of thought. For example, mathematicians know that coding in three variables is more efficient than binary coding, but for historical and technological reasons binary logic has come to dominate digital computers, and most people now believe that this is so because binary is the best. If there were no constraints, the most efficient coding is $e = 2.71828...$ which is a well known result of mathematical logic.[71]

It must be acknowledged that ancient seers and philosophers saw reality in terms of triads. At the deepest level, these triads came into play due to a division of reality into subject, object, and the medium that makes for an interaction between the two.

Nature chooses optimality: the principle of least action is one way to explain dynamics in classical and quantum physics, and it is central in the understanding of evolutionary biology. Therefore, information optimality should figure in the foundations of theory.

Our maps of reality are constructed based on information received through the senses by the mind, which then associates numbers with this data. In this construction, memory plays a crucial role, for information makes sense only in relation to the past and expectations about the future. Information is primary, and conceptual schemes such as space as a container are secondary constructs of the mind.

Space is a map of the received sensations into bins, and the optimality of nature compels the notion that its measure, that is dimensionality, should be e. The mathematics of noninteger spaces and fractals provides a way to investigate properties of noninteger dimensionality.[72] Fractals make it possible to imagine similarity in structure across different scales.

[71] Kak, S. Information theory and dimensionality of space. Scientific Reports (2020)

[72] Stillinger, F.H. Axiomatic basis for spaces with noninteger dimension. J. Math. Phys. 18 (1977)

The Big Bang model

In the Big Bang model, the universe began in a singularity (initial state of extremely high density and high temperature) that exploded and expanded exponentially in an epoch that lasted between 10^{-36} and 10^{-33} seconds. After this the universe continued to expand at a much slower pace.

Without going into how energy, time, and space emerged, it can explain the cosmic microwave background (CMB) radiation and the large-scale structure of the universe with its galaxies and bigger clusters.

The model explains why stars that are farther away are receding faster and by analyzing the measurements of the expansion of the most distant objects, it has been concluded that the Big Bang singularity occurred around 13.8 billion years ago. Newer measurements on supernovae show that the expansion of the universe is accelerating. Stars within the galaxies also seem to move faster than can be explained by gravitation. To explain these findings, dark energy and dark matter have been postulated and together they are supposed to be around 96 percent of the observable universe. Given that interstellar dust constitutes an additional 3.5 percent, this means that our theories about the cosmos are based on the understanding of merely 0.5 percent of the universe. Furthermore there is no direct evidence in support of either dark matter or dark energy. Where do we go from here? If 96 percent of the universe is missing, perhaps our theories are hopelessly wrong.

The measure of the expansion rate of space is the Hubble constant. But there is divergence between the expansion rates obtained from the early (captured by CMB data) and the late (considering the receding stars and galaxies) aspects of the universe. The Hubble constant is measured using the CMB data by measuring the density variations of matter caused by reverberating sound waves that are imprinted on the CMB as temperature variations. Calculating the constant for the late universe requires measuring both the velocities and distances of far-off galaxies. For distances which are too large to measure using parallax, astronomers use standard candles.

There are several types of standard candle objects for which we can predict luminosity from some other measurement, and two of the most important ones are Cepheid variable stars and type 1A supernovae. Cepheids variable stars pulsate with a frequency that is proportional to their absolute magnitude, and they are used to measure the distance of galaxies out to about 50 Mpc, whereas type 1A supernovae can be seen at distances from about 1 Mpc to over 1,000 Mpc, where a parsec is approximately equal to 3.26 light-years. The methods require climbing a distance ladder by stacking up different measuring methods tied to different kinds of bright objects. The expansion rate based on early universe is about 67 km s-1 Mpc-1, whereas a late universe estimate is about 74 km s-1 Mpc-1. Errors of observation have been ruled out and many have said that it represents a crisis in cosmology.

The discrepancy gets resolved if the early universe data is associated with the physical universe of dimensionality *e* whereas the late universe data is based on 3-dimensionality (for we fit data from stars into the three-dimensional maps of our models). The expansion is measured in relation to its direction which implies a discrepancy equal to e/3=0.9060, for dimensionality along each direction should be one-third of the aggregate value. This number is very close to the divergence of 67/74=0.9054 from the experimental data.

What is space?

The beginnings of the universe and the nature of space are connected to deep questions of physics and philosophy. Where do time, energy and matter come from? Since intuition of space is also related to memory, the larger question includes the origin of consciousness.

The Western tradition takes space to be absolute and independent of observers in it, or in other words it is seen as a container. Aristotle, who was the first to use the word "physics" in the sense that it is understood now, conflated change in biological and physical domains. He defined motion as the actuality of a potentiality, which is fine in the "motion" of a living organism, but wrong otherwise. Defining motion this way requires the assumption of an absolute frame and other imaginary schema. He gave examples of four types of change, namely change in substance, in quality, in quantity and in place, without providing logical bases for this assertion.

According to Aristotle, the sun, the moon, planets and stars are embedded in perfectly concentric crystal spheres that rotate at fixed rates. The celestial spheres are made up of the element ether which supports uniform circular motion. He took the terrestrial objects to be composed of four other elements that rise or fall. The earth, the heaviest element, and water, fall toward the center of the universe; hence the earth and the oceans constitute our planet. At the opposite end, the lightest elements, air and fire, rise and move away from the center.

With the rise of Christianity, Europe was cut off from its past and many Greek scientific texts were lost or preserved only in translation. Aristotle's Physica and De Caelo (On the Heavens) were translated from Arabic to Latin in the twelfth century. Soon after, Thomas Aquinas in his Summa Theologica (1265–1274) reconciled Aristotle's ideas to the demands of Christian dogma.

This explains why the challenge to the idea of the earth being the center of the solar system (or the universe) was such a big thing in Europe. In 1600, Giordano Bruno was burnt at stake for his heresy against the geocentric model. Galileo was tried by the Inquisition, found vehemently suspect of heresy, and forced to recant. He spent the rest of his life under house arrest, before dying in 1642. The mathematical theory of gravitation that was to emerge later in the seventeenth century provided an explanation for the motions of astronomical objects, but it did not alter the basic picture.

In the observer-centric Indian physics that goes back to Kaṇāda of about 600 BCE, physical laws must be based only on nature of the physical universe and the experience of time and space is a consequence of the relation between the observer and the world being observed.

In this view, all material properties are a consequence of substances, their characteristics, and their motions, together with interactions, and as far as the observers are concerned, they have intuitions related to universality (invariances), particularities (depending on the nature of the observer), and the interaction between the observer and the system. Clearly, this view privileges intuition over any pre-conceived ideas about the ontology of space and time.

e–Dimensionality and gravity

Physicists have been unable to unify gravitation with the other known forces of electromagnetism, and the weak and the strong nuclear forces that are important in the study of matter. Has this happened because the explanation for gravity is being sought in the wrong place?

Consider the idea of dimensionality afresh. Dimension 0 is a point, dimension 1 is a line, dimension 2 is a plane, and dimension 3 is a solid, and these are frames in which we see their contents, and objects in them stay where they are. What about noninteger dimensionality? Superficially, it means that somehow the space is less dense than that of the immediately higher dimension. But space cannot be like a sponge, with holes in it that reduce its density, although that is how we view fractals. As a fundamental aspect of reality, space should have the same characteristics at all scales.

Let us consider the dimensionality potential of an object. The greater the separation between two points, the greater is the dimensionality attraction, as in the example of d =0.7 where objects will tend to get closer to each other by this factor. Although the potential p(r) will increase with distance r, it should be divided by the surface area, A, of the equipotential surface around it.

This change in the gravitational force can explain the recent phase of the acceleration of the expansion of the universe.

Fractals have dimensionality that is noninteger and structure that repeats across different scales. Fractals are found all over nature: there are self-similar structures in the brain, in plants, and at cosmic scales. Three examples of fractals are the Whirlpool Galaxy, the Nautilus shell, and Romanesco broccoli shown in the next figures.

M51: The Whirlpool Galaxy

The old idea of space as something static must be wrong. If space at integer dimensionality is static, it is not so at noninteger values. Taken broadly, dimensionality of less than 3 makes space shrink in a dynamic way, which we see as the source of gravitational attraction. If gravity is a property of space, it solves a puzzle for which science has had no answer until now.

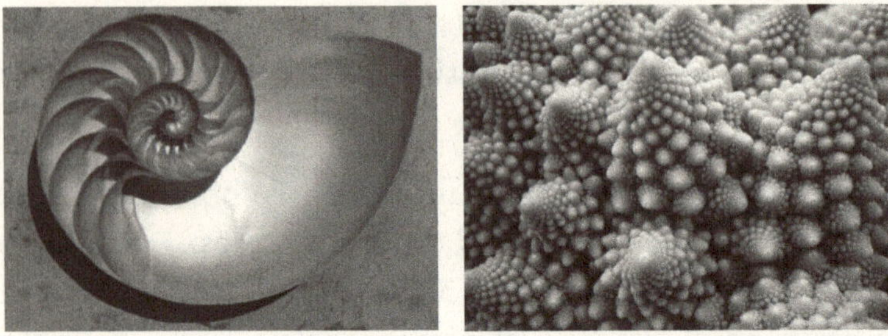

The Nautilus shell (left) and Romanesco broccoli (right)

Space has potential and energy, and we can see that in the evolution as the universe goes from the dimensionality of zero to the optimum value of *e* and we can use this lens to examine cosmology.

The sponge-view is one way of looking at space that is used in mathematical models; the dual view is that dynamics itself is an expression of this nature. Such disparate views are harmonized by the principle of complementarity, which is one of the deepest philosophical ideas in science. In other words, noninteger dimensionality and gravitational attraction are two complementary ways of speaking of a deeper reality.

The postulation of dimensional energy and evolution of zero-dimensional singularity into higher dimensions provides an alternative to the current model of cosmology. As in the established theory, the expansion goes through different stages: (i) a very rapid initial expansion at a nearly instantaneous rate; (ii) an inverse-square law attraction mode with two sub-phases (radiation-dominant and matter-dominant) where this attraction becomes increasingly larger which slows down the expansion from its initial phase; (iii) accelerated expansion as the attraction force declines and gravitation holds steady, and (iv) decelerating expansion followed eventually by contraction of the universe.[73]

Looking at expansion of the universe as a property of the dimensionality of space obviates the need to postulate dark energy as being responsible for the accelerated expansion, but it indicates that gravity has declined by 20% from its peak.

In mainstream cosmology, black holes are seen as emerging out of one of the following three scenarios: from the core-collapse of a sufficiently massive star, from the direct collapse of either a massive star or gas cloud, or from the collision of one compact object, like a neutron star, with another. In this view, it takes a long time in the early universe for density deviations to build up so that gravitational collapse can take place, which in turn leads to the formation of stars. It is only much later, that very large stars can merge and form supermassive black holes that we see today.

The new theory presents a totally different picture. Since the universe evolves with increasing dimensionality, one may conclude that primordial black holes, filaments, barred, and disk galaxies will be common in the early universe.[74] For $d < 2$, there would be point, linear or bar structures and as the universe evolved into the $d > 2$ phase these would be seen as primordial black holes, barred spiral galaxies (as in NGC 1672 or IC 5201) together with double-barred galaxies.

The recent discovery of Galaxy DLA0817g, nicknamed the Wolfe Disk, made with the Atacama Large Millimeter/

[73] Kak, S. Evolutionary stages in a noninteger dimensional universe. Indian Journal of Physics, 97, 3041–3045 (2023)

[74] Kak, S. Filaments and spirals in a noninteger dimensional universe. Indian Journal of Physics, 99, 1569-1574 (2025)

submillimeter Array (ALMA) of a massive rotating disk galaxy, seen when the Universe was only ten percent of its current age, challenges the traditional models of galaxy formation. This, the most distant rotating disk galaxy ever observed, is contrary to most galaxy formation scenarios in which galaxies only start to show a well-formed disk around 6 billion years after the Big Bang. More recently, the galaxies called SPT-S J041839-4751.9 and BRI 1335-0417 only 1.4 billion years after the Big Bang have also been revealed to have a spiral structure.

Since the notion of dimension applies to physical reality at all conceivable scales, it should be relevant to biology and cognitive science and for the study of anomalous mechanical properties of materials. In the case of the genetic code, it provides a rationale for why the groupings of codons per amino acid range non-uniformly from 1 to 6, which solves a long-standing puzzle of biology.[75] It also explains why the B-form DNA is the most common of its three forms, which include the A- and the Z-forms.[76]

[75] Kak, S. Self-similarity and the maximum entropy principle in the genetic code. Theory in Biosciences 142, 205- 210 (2023)

[76] Kak, S. Fractal dimensionality of a coiled helical coil. Journal of Bioinformatics and Computational Biology 23 (2025)

9

Into the Future

The challenges related to colonization of Mars and the planetary moons are both external and internal. The external challenges relate to science and technology that are required to make the physical habitats possible, but equally important are the psychological challenges that inhabitants of these far-flung colonies will face.

It is one thing to go on a short-term visit to a remote space location – even if it is several months or a year or two – for one can always dream of returning back to one's home and family and friends, but it is quite another of having to leave Earth permanently and live in a small colony of a few thousand and live there for the rest of one's life. This will require special mental and emotional strength that few people are capable of, and it will also be a great problem for children who are born in these remote places.

The questions of meaning of life are even more urgent in space than they are on Earth and in the technology governed post-modern world, they are already a cause of loneliness and depression. We need a better understanding of the attitudes and beliefs of those who choose to live alone at remote places on Earth, like those who leave their homes and live in the caves in the Himalayas. Therefore, questions of philosophy and spirituality will be significant in the choosing of the residents of Mars and interplanetary moons.

The Mars Society is a nonprofit organization, founded in 1998, that advocates for human exploration and colonization of the planet. It generates interest in the program through lobbying, organizing the annual International Mars Society Convention and operating Mars analog research stations to study the medical and psychosocial effects of long-term space missions. Explore Mars is another organization that is committed to achieving human exploration of Mars by the 2030s. The Planetary Society is also working towards exploration of the planets.

Elon Musk has said that humans will be on Mars by 2029 and there would be a million-person city there in twenty or thirty years. We are not that optimistic, and possibly in that time window we would be looking at best at settlements of a few hundred, although the founder of the Mars Society thinks a city of 50,000 is realistic.

Given the current crisis of cosmology that was described in an earlier chapter, our understanding of the universe may be in for a major revision in the coming decades. Ideas such as that the universe is quantum-like, time is an illusion of the mind and locally determined, and the boundaries of the universe are beyond space and time may become more mainstreamed. There are theories that the universe is a black hole, and in such a scenario there could even be black holes within black holes. One could search the boundaries beyond the beginning of the universe, beyond the Alpha, all the way to the Omega, beyond the black hole that the universe may turn out to be.

On the other hand, the idea that space is not three-dimensional but rather *e*-dimensional as one of us has shown[77] may have implications for the space exploration enterprise. It may even be possible to exploit it in new ways that help in the space travel process.

Much remains to be done to unify the laws of quantum mechanics and the laws of general relativity (GR). Such unification is impossible if we treat the world as a purely physical system. The role of the mind must be brought in from the get-go, and maybe the mind has played a paramount role in evolution.

We must become the stewards of all life on Earth. The future humans will belong to the world ecological system as one dynamic, evolutionary unit. This will go beyond the current limited vision of integrating life with technology and AI.

We must address macro evolution, AI and life to address the challenges humans will confront in the planetary colonies. The aloneness of the existence in these remote outposts will bring up questions of the meaning of life and those of beginnings and endings.

---

[77] Kak, S. Information theory and dimensionality of space. Scientific Reports (2020); Kak, S. and Kafatos, M. Black holes, disk structures, and cosmological implications in *e*-dimensional space. Physics Essays 35, 345–355 (2022)

The one without beginning and end is Being itself. We may call it the One, and it projects into a multiplicity. The universe is a rich kaleidoscope of objects, patterns, colors, sounds, patterns of vibration, ways of interactions and of course countless life forms. As we try to make sense of limited "things", they make up the whole quite like the case of patterns that characterize a rich carpet.

The genetic code is remarkably simple, yet it is responsible for a vast array of information defining the individual and the species. Given that the elements of the code have been found on asteroids, one may assume that life across galaxies may have a structure like what we have on Earth.

In many ways, what is unique to different species and separates different cultures is the degree of their complexity. We try to make sense of patterns, forgetting that patterns are just indicators of biological complexity and evolution.

The many elementary particles make up many molecules, the many molecules make up cells, the many cells make up organisms, the many organisms make up ecosystems, the many ecosystems make up life on Earth and maybe beyond.

It appears that complexity in evolution, and in the many neurons in the brain is driven by consciousness. In such a sense, consciousness is the driver of everything and being and consciousness are inseparable.

Science and spirituality are together; they are not separate. What appears as separation is a mere set of relationships that are based on the solid foundation of existence. When we finally understand that the vast number of interplays is the state of becoming that arises from existence, then we will move closer to what the ancient thinkers have been urging us to undertake in our quest. This is true philosophy, being "friends of wisdom" as the best of all minds have always declared.

In examining the foundations of the physical universe, we examine the "bricks" on which physical structures, space and time or in the language of relativistic physics, spacetime. Space is somewhat easier to understand, once we have more than one object, more than one atom etc., space is what exists between objects. We also mentioned *e*-space which is likely the *real* space. In a sense, space is objective and can be experienced as such.

Time is a bit trickier. We can only speak of its intervals, the lifetime of an object, and even the lifetime of the universe. Cosmic time is certainly different than the lifetime of a galaxy, the lifetime of a star (which depends on its mass; hence it follows stellar evolution), the lifetime of a planet and, last but not least, the time involved in biological evolution. "Time" is the ultimate property associated with objects, and in a sense it is more "subjective" than space[78].

In the relativity theory, spacetime is taken to be flat which is a limiting case of general relativity that allows curved spacetime. In both, space and time are affected by the relative motion between the observer and the object. The lengths along the direction of motion, say within a spacecraft, are shrunk, and time is slowed down compared to a stationary observer. But inside the spaceship things

[78] Kak, S. and Kafatos, A.M. Dimensionality of time and the information paradox. Physics Essays, 38 (2025)

remain normal. It is the outside, stationary observer that believes relativity affects space and time for the moving spaceship.

Then there are certain peculiarities associated with time, such as the twin paradox. Suppose we have twins, one remains on Earth, the other travels in space with great speed, say approaching the speed of light. The one who goes out in space and then turns around and flies back to Earth to meet her twin will find that she has aged much less than the twin who stayed back.

If a spacecraft travels to a black hole the distant observer will see it coming closer and closer to the event horizon (the boundary of the black hole) but not be able to cross it. If the spacecraft were sending signals back to Earth, the observer would see the messages redshift and slow down as it neared the event horizon, eventually stretching out infinitely. From the observer's perspective, the spacecraft will never quite reach it—it would be seen at the edge forever. But from the spacecraft's perspective, the movement will go forward normally.

Once the spacecraft passes the event horizon, there's no escape. As it approaches the singularity, tidal forces grow infinite, and it will be crushed by them.

If our universe exists within a black hole then there are other universes beyond ours that exist within other black holes. In this view, space and time are of infinite duration that lie within a consciousness of infinite potential.

Quantum mechanics has shown us that the universe as experienced and described by us exists in the mind. This is because real properties of objects, starting with quanta, are derivatives of the observing mind and are not existing devoid of observation.

Nevertheless, there is objectivity of quantum states, where this term implies the existence of observers (they don't have to be conscious observers, however, if mechanical or electromagnetic, they should be able to distinguish "up" from "down"). The ultimate binary choice is "yes" or "no". This choice is, again, based on some sort of observation.

One may assert that philosophically *Universal Consciousness* is the ultimate reality, and it is the constant companion of *One Being.* While the visions of the One are many, the Subject is One. This understanding is very much *within science*, and it is based on quantum mechanics and it is consistent with Vedanta.

Thus, there is a single Reality, quite as the ancient philosophers and modern scientists have argued, which is the natural outcome of the search for the one scientific truth. This vision will be central in the unfolding of Project Omega.

Bibliography

Barron, A.B., Klein, C. What insects can tell us about the origins of consciousness. Proc. Natl. Acad. Sci. U. S. A., 113, 4900-4908 (2016)

Basham, A.L. The Wonder That Was India, Sidgwick and Jackson (2000)

Baum D.A. The origin and early evolution of life in chemical composition space. J. Theor. Biol.456, 295–304 (2018)

Bhatia, G.K., Sahijpal, S. Thermal evolution of trans-Neptunian objects, icy satellites, and minor icy planets in the early solar system. Meteoritics & Planetary Science. 52 (12): 2470–2490 (2017)

Birch, J., Schnell, A.K., Clayton, N.S. Dimensions of animal consciousness, Trends in Cognitive Sciences 24, 789-801 (2020)

Booth, L. N. et al. The aging epigenome. Molecular Cell 62, 728 – 744 (2016)

Brannen, P. The Ends of the World: Volcanic Apocalypses, Lethal Oceans, and Our Quest to Understand Earth's Past Mass Extinctions. Ecco (2018)

Brookes, J.C. Quantum effects in biology: golden rule in enzymes, olfaction, photosynthesis and magnetodetection. Proceedings of the Royal Society A: Mathematical, Physical and Engineering Sciences. 473 (2017).

Cabrol, N.A. The Secret Life of the Universe. Scribner (2024)

Carroll, M. Envisioning Exoplanets: Searching for Life in the Galaxy. Smithsonian Books (2020)

Carroll, S. The Biggest Ideas in the Universe: Space, Time, and Motion. Dutton (2022)

Chopra, D., Kafatos, M.C. You Are the Universe. Harmony, Random House (2017)

Costantin, L. et al. A Milky Way-like barred spiral galaxy at a redshift of 3. Nature. 623 (7987), 499–501 (2023)

Dimmitt, C., van Buitenen, J.A.B. Classical Hindu Mythology: A Reader in the Sanskrit Puranas. Temple University Press (2012)

Emlen, D., Zimmer, C. Evolution: Making Sense of Life. W.H. Freeman (2019)

Feldman, J. M., and Lynch, J. G. Self-generated validity and other effects of measurement on belief, attitude, intention, and behavior. J. Appl. Psychol. 73, 421–435 (1988)

Futuyma, D.J., Kirkpatrick, M. Evolution. Sinauer Associates (2017)

Garbe, R. On the voluntary trance of Indian fakirs. The Monist 10 (4), 481–500 (1900)

Gard, T., Noggle, J.J., Park, C.L., Vago, D.R., Wilson A. Potential self-regulatory mechanisms of yoga for psychological health. Front Hum Neurosci. 30, 770 (2014)

Garvin J.B.et al. Revealing the Mysteries of Venus: The DAVINCI Mission. Planet. Sci. J. 3, 117 (2022)

Goh, B.H., Tong, E.S., Pusparajah, P. Quantum Biology: Does quantum physics hold the key to revolutionizing medicine? Progress in Drug Discovery & Biomedical Science 3 (2020)

Goldstein, B.R. Eratosthenes on the "measurement" of the Earth. Historia Mathematica 11, 411-416 (1984)

Greene, B. The Fabric of the Cosmos: Space, Time, and the Texture of Reality. Vintage (2005)

Greene, B. Until the End of Time: Mind, Matter, and Our Search for Meaning in an Evolving Universe. Vintage (2021)

Grossberg, S. The complementary brain: unifying brain dynamics and modularity. Trends Cogn. Sci. 4, 233–246 (2000)

Guo, J. et al. Aging and aging-related diseases: from molecular mechanisms to interventions and treatments. Sig Transduct Target Ther 7, 391 (2022)

Hadamard, J., The Psychology of Invention in the Mathematical Field, Dover, New York (1954)

Hall, K.T. Placebos. The MIT Press (2022)

Harvey, R. The Solar System: Exploring the Sun, Planets and Their Moons. Amber Books (2022)

Hawking, S. A Brief History of Time. Bantam (2011)

Hölzel, B.K. et al. Mindfulness practice leads to increases in regional brain gray matter density. Psychiatry Res. ;191, 36-43 (2011)

Howell S.M., Pappalardo R.T. NASA's Europa Clipper—A Mission to a Potentially Habitable Ocean World. Nat. Commun. 11, 1311 (2020)

Kafatos, M. and Nadeau, R. The Conscious Universe: Parts and Wholes in Physical Reality, Springer (2000)

Kafatos, M.C. and Kak, S. Veiled nonlocality, cosmic censorship, and local observations. Physics Essays 28, 182-187 (2015)

Kafatos, M.C. and Kato, G.C. Sheaf theoretic formulation for consciousness and qualia and relationship to the idealism of non-dual philosophies. Progress in Biophysics and Molecular Biology131, 242-250 (2017)

Kafatos, M.C. Science, Reality & Everyday Life. Prisma Press, Auroville, India (2024)

Kafatos, M.C., Banerji, D., and Struppa, D. (editors). Quantum and Consciousness Revisited. DK Publisher (2024)

Kak, S. and Kafatos, M. Black holes, disk structures, and cosmological implications in e-dimensional space. Physics Essays 35, 345–355 (2022)

Kak, S. The structure of the Ṛgveda. Indian Journal of History of Science 28, 71-79 (1993)

Kak, S. and Kafatos, M. Dimensionality of time and the information paradox. Physics Essays, 38 (2025)

Kak, S. Information theory and dimensionality of space. Scientific Reports (2020)

Kak, S. No-go theorems on machine consciousness. Journal of Artificial Intelligence and Consciousness 10, 237-247 (2023)

Kak, S. Self-similarity and the maximum entropy principle in the genetic code. Theory in Biosciences 142, 205- 210 (2023)

Kak, S. Evolutionary stages in a noninteger dimensional universe. Indian Journal of Physics, 97, 3041–3045 (2023)

Kak, S., The Age of Artificial Intelligence. Garuda (2024)

Kak, S. Time, consciousness, and lifespan. Journal of Artificial Intelligence and Consciousness, 12 (2025)

Kak, S. Filaments and spirals in a noninteger dimensional universe. Indian Journal of Physics, 99, 1569-1574 (2025)

Kak, S. Fractal dimensionality of a coiled helical coil. Journal of Bioinformatics and Computational Biology 23 (2025)

Koshino, K. and Shimizu, A. Quantum Zeno Effect by General Measurements. Physics Report, 412 (2005)

Laub, D. Thanatos and massive psychic trauma: the impact of the death instinct on knowing, remembering, and forgetting. Journal of the American Psychoanalytic Association 51(2), 433-64 (2003)

Limbach, M.A. et al. The MIRI Exoplanets Orbiting White Dwarfs (MEOW) Survey: Mid-infrared Excess Reveals a Giant Planet Candidate around a Nearby White Dwarf. ApJL 973 L11 (2024)

McGarey, P. et al., How to Deploy a 10-km Interferometric Radio Telescope on the Moon with Just Four Tethered Robots. 2022 IEEE Aerospace Conference (AERO), Big Sky, MT, USA, 2022.

Merker B. Consciousness without a cerebral cortex: a challenge for neuroscience and medicine. Behav Brain Sci. 30(1), 63-81 (2007)

Misra, B. and Sudarshan, E.C.G. The Zeno's Paradox in Quantum Theory. J. Math. Phys 18, 756-763 (1977)

Möbus, T. and Wolf, M.M. Quantum Zeno effect generalized. J. of Math. Phys. 60 (2019)

Murphy, M. The Future of the Body: Explorations Into the Further Evolution Of Human Nature. Penguin (1993)

Nielsen, M. and Chuang, I. Quantum Computation and Quantum information. Cambridge University Press (2010)

Ou, H.L., Schumacher, B. DNA damage responses and p53 in the aging process. Blood 131 (5), 488–495 (2018)

Patil, Y. S., Chakram, S., Vengalattore, M. Measurement-induced localization of an ultracold lattice gas. Physical Review Letters 115, 140402 (2015)

Penrose, R. The Emperor's New Mind: Concerning Computers, Minds and the Laws of Physics. Oxford University Press (1989)

Penrose, R. The Road to Reality. Vintage (2007)

Polidan, R. et al. FarView: An in-situ manufactured lunar far side radio array concept for 21-cm Dark Ages cosmology. Advances in Space Research 74, 528-546 (2024)

Puumala, M.M., Sivula, O., Lehto, K. Moving to Mars: The Feasibility and Desirability of Mars Settlements. Space Policy 66 (2023)

Rovelli, C. Seven Brief Lessons on Physics. Riverhead Books (2016)

Rowland, Lucie E.; et al. REBELS-25: Discovery of a dynamically cold disc galaxy at z = 7.31. Monthly Notices of the Royal Astro. Society 535 (3), 2068–2091 (2024)

Schenk, P.M. et al. Enceladus and the Icy Moons of Saturn. University of Arizona Press (2018)

Schrödinger, W. What is Life? and Mind and Matter. Cambridge University Press (1967)

Shobana, R. et al. Effect of long-term yoga training on autonomic function among the healthy adults. J. Family Med. Prim. Care 11, 3471-3475 (2022)

Slocombe, L., Al-Khalili, J. S.; Sacchi, M. Quantum and classical effects in DNA point mutations: Watson-Crick tautomerism in AT and GC base pairs. Physical Chemistry Chemical Physics 23, 4141–4150 (2021)

Spilker L. Cassini-Huygens' Exploration of the Saturn System: 13 Years of Discovery. Science. (2019)

Stein, M.A. (tr.), Kalhana's Rajatarangini. A Chronicle of the Kings of Kasmir, Vol. 1. Delhi: Motilal Banarsidass (1961). Reprint of the 1900 edition.

Streeter, C.C. et al. Effects of yoga on the autonomic nervous system, gamma-aminobutyric-acid, and allostasis in epilepsy, depression, and post-traumatic stress disorder. Medical Hypotheses 78, 571-579 (2012)

Summers, M. Trefil, J. Exoplanets: Diamond Worlds, Super Earths, Pulsar Planets, and the New Search for Life beyond Our Solar System. Smithsonian Books (2018)

Tsai, S.Y. et al. Meditation Effects on the Control of Involuntary Contingent Reorienting Revealed With Electroencephalographic and Behavioral Evidence. Front Integr Neurosci. 12, 17 (2018)

Venkatesananda, S. Vasistha's Yoga. SUNY Press (2010)

Vijg, J. From DNA damage to mutations: All roads lead to aging. Ageing Research Reviews 68, 101316 (2021)

Von Braun, W. Project Mars. Apogee Books (2006).

Wang, B. What Disaster Happened? 2200, The World Only Has 250 Million People. Next Big Future (2023)

Wilson, E.O. The Future of Life. Vintage Books (2002)

Wilson, H.H. (tr.), The Vishnu Purana, (1840)

Zubrin, R. The Mars Dream Is Back — Here's How to Make It Actually Happen. The New Atlantis (2025)

About the Authors

Subhash Kak is Regents Professor at Oklahoma State University in Stillwater and a Distinguished Academic Scholar at Chapman University. He has authored thirty books, of which the most recent ones are *Eternal Bharat: Truth, Meaning, and Beauty* and *The Age of Artificial Intelligence*. Since 2018, he has been a member of the Indian Prime Minister's Science, Technology, and Innovation Advisory Council.

Menas C. Kafatos is Fletcher Jones Endowed Professorship Computational Physics and Director, Institute for Earth, Computing, Human & Observing (ECHO) at Chapman University. He is the author and editor of 25 books including the New York Times Bestseller *You are the Universe* coauthored with Deepak Chopra. He works or collaborates with many institutions in the U.S., in Korea, such as Korea University and Ewha Womans University, as well as the Physics Department in the University of Athens and the Academy of Athens.